AN INTRODUCTION TO NONHARMONIC FOURIER SERIES

REVISED FIRST EDITION

An Introduction to Nonharmonic Fourier Series

Revised First Edition

Robert M. Young
James F. Clark Professor of Mathematics
Department of Mathematics
Oberlin College
Oberlin, Ohio

ACADEMIC PRESS
A Harcourt Science and Technology Company

San Diego San Francisco New York Boston
London Sydney Tokyo

This book is printed on acid-free paper. ∞

ACADEMIC PRESS
A Harcourt Science and Technology Company
525 B Street, Suite 1900, San Diego, CA 92101-4495, USA
http://www.academicpress.com

Academic Press
Harcourt Place, 32 Jamestown Road, London NW1 7BY, UK
http://www.academicpress.com

Library of Congress Catalog Card Number: 00-104370

International Standard Book Number: 0-12-772955-0

Printed in the United States of America

01 02 03 04 05 MB 9 8 7 6 5 4 3 2 1

To Linda

CONTENTS

3

THE COMPLETENESS OF SETS OF COMPLEX EXPONENTIALS

4

INTERPOLATION AND BASES IN HILBERT SPACE

PREFACE TO THE REVISED FIRST EDITION

When *An Introduction to Nonharmonic Fourier Series* first appeared, there was little reason to believe that the theory of Riesz bases and frames in Hilbert space would one day play so prominent a role in both pure and applied harmonic analysis. The "wavelets revolution" of the past fifteen years helped change that. Now, the three classic treatises in the field — arguably those by Daubechies [1992], Meyer [1992], and Mallat [1999] — all attest to the fact that research in frame theory flourishes. Moreover, in the more specialized area of nonharmonic Fourier series, the celebrated work of Avdonin and Ivanov [1995] provides ample evidence that Riesz bases of complex exponentials remain of central importance in the theory of control.

I am therefore deeply indebted to Bob Ross and the editorial staff of Academic Press for the opportunity to bring *An Introduction to Nonharmonic Fourier Series* up to date. To maintain the continuity of the original text, most of the new material has been incorporated in a greatly expanded set of Notes and References. In this way, the book remains true to its original goal — to provide an elementary introduction to a rich and multifaceted field, not an exhaustive account of all that is known.

PREFACE TO THE FIRST EDITION

The theory of nonharmonic Fourier series is concerned with the completeness and expansion properties of sets of complex exponentials $\{e^{i\lambda_n t}\}$ in $L^p[-\pi, \pi]$. Its origins, which are classical in spirit, lie in the celebrated works of Paley and Wiener [1934] and Levinson [1940]. In recent years, in response to the development of functional analysis and, in particular, to the growing interest in bases in Banach spaces, research in the area has flourished. New approaches to old problems have led to important advances in the theory.

This book is an account of both the classical and the modern theories. Its underlying theme is the elegant interplay among the various parts of analysis. The catalyst in the present case is the Fourier transform, through which the classical Banach spaces are mapped into spaces of entire functions. In this way, problems in one domain can be examined via their transform image in the other.

The book is designed primarily for the graduate student or mathematician who is approaching the subject for the first time. Its aim as such is to provide a unified and self-contained introduction to a multifaceted field, not an exhaustive account of all that is known. Accordingly, the first half of the book presents an elementary introduction to the theory of bases in Banach spaces and the theory of entire functions of exponential type. At the same time, an extensive set of notes touches on more advanced topics, indicates directions in which the theory can be extended, and should prove useful to both specialists and nonspecialists alike. Much of the material appears in book form for the first time.

The only prerequisites are a working knowledge of real and complex analysis, together with the elements of functional analysis. By that I mean roughly what is contained in Rudin [1966]. On occasion, when more advanced tools of analysis are required, appropriate references are given. Apart from this, the work is essentially self-contained, and it can serve as a textbook for a course at the second- or third-year graduate level.

The problems, which are of varying degrees of difficulty, are an integral part of the text. Some are routine applications of the theory, while others are important ancillary results — these are usually accompanied by an indication of the solution and an appropriate reference to the literature.

A word about notation: Theorem 2.3 refers to Theorem 3 of Chapter 2. The labeling of all other results is self-explanatory.

I am deeply indebted to Doug Dickson and Paul Muhly for their careful reading of the manuscript and for their sharp criticism and advice. I owe immeasurable thanks to Linda Miller, who proofread the entire book more times than I could possibly have hoped.

ROBERT M. YOUNG

1

BASES IN BANACH SPACES

1 SCHAUDER BASES

Let X be an infinite-dimensional Banach space over the field of real or complex numbers. When viewed as a vector space, X is known to possess a *Hamel basis* — a linearly independent subset of X that spans the entire space. Unfortunately, such bases cannot in general be constructed, their very existence depending on the axiom of choice, and their usefulness is therefore severely limited. Of far greater importance and applicability in analysis is the notion of a basis first introduced by Schauder [1927].

Definition. *A sequence of vectors $\{x_1, x_2, x_3, \ldots\}$ in an infinite-dimensional Banach space X is said to be a* **Schauder basis** *for X if to each vector x in the space there corresponds a unique sequence of scalars $\{c_1, c_2, c_3, \ldots\}$ such that*

$$x = \sum_{n=1}^{\infty} c_n x_n.$$

The convergence of the series is understood to be with respect to the strong (norm) topology of X; in other words,

$$\left\| x - \sum_{i=1}^{n} c_i x_i \right\| \to 0 \quad as \quad n \to \infty.$$

Henceforth, the term *basis* for an infinite-dimensional Banach space will always mean a Schauder basis.

Example. The Banach space $l^p (1 \leqq p < \infty)$ consists, by definition, of all infinite sequences of scalars $c = \{c_1, c_2, c_3, \ldots\}$ such that $\|c\|_p = (\sum_{n=1}^{\infty} |c_n|^p)^{1/p} < \infty$. The vector operations are coordinatewise. In each of these spaces, the "natural basis" $\{e_1, e_2, e_3, \ldots\}$, where

$$e_n = (0, 0, \ldots, 0, 1, 0, \ldots),$$

and the 1 appears in the nth position, is easily seen to be a Schauder basis. If $c = \{c_n\}$ is in l^p, then the obvious expansion $c = \sum_{n=1}^{\infty} c_n e_n$ is valid.

It is clear that a Banach space with a basis must be *separable*. Reason: If $\{x_n\}$ is a basis for X, then the set of all finite linear combinations $\sum c_n x_n$, where the c_n are *rational* scalars, is countable and dense in X. It follows, for example, that since l^∞ is not separable, it cannot possess a basis.

The "basis problem"—whether or not every separable Banach space has a basis—was raised by Banach [1932] and remained until recently one of the outstanding unsolved problems of functional analysis. The question was finally settled by Per Enflo [1973], who constructed an example of a separable Banach space having no basis. The negative answer to the basis problem is perhaps surprising in light of the fact that bases are now known for almost all the familiar examples of infinite-dimensional separable Banach spaces.

PROBLEMS

1. Prove that every vector space has a Hamel basis.

2. Prove that every Hamel basis for a given vector space has the same number of elements. This number is called the (linear) **dimension** of the space.

3. Show that a Hamel basis for an infinite-dimensional Banach space is uncountable.

4. Show that the dimension of l^∞ is equal to c. (*Hint*: Show that the set $\{(1, r, r^2, \ldots) : 0 < r < 1\}$ is linearly independent.)

5. Let X be an infinite-dimensional Banach space.
   (a) Prove that $\dim X \geqq c$. (*Hint*: Show that there is a vector space isomorphism between l^∞ and a subspace of X.)
   (b) Prove that if X is separable, then $\dim X = c$.

6. The Banach space c_0 consists of all infinite sequences of scalars which converge to zero (with the l^∞ norm). Show that the natural basis is a Schauder basis for c_0.

7. Exhibit a Schauder basis for the Banach space c consisting of all convergent sequences of scalars (with the l^∞ norm).

8. An infinite series $\sum x_n$ in a Banach space X is said to be **unconditionally convergent** if every arrangement of its terms converges to the same element. It is said to be **absolutely convergent** if the series $\sum \|x_n\|$ is convergent. Show that every absolutely convergent series in X is unconditionally convergent. What about the converse?

9. A basis $\{x_n\}$ for a Banach space X is said to be **unconditional (absolute)** if every convergent series of the form $\sum c_n x_n$ is unconditionally (absolutely) convergent.
   (a) Show that the natural basis is unconditional for the spaces l^p, $1 \leqq p < \infty$, and c_0. Show also that it is absolute for l^p only when $p = 1$. Is it absolute for c_0?
   (b) Show that the sequence of vectors

$$(1, 0, 0, 0, \ldots), \quad (1, 1, 0, 0, \ldots), \quad (1, 1, 1, 0, \ldots), \quad \ldots$$

forms a basis for c_0 which is not unconditional.

2 SCHAUDER'S BASIS FOR $C[a, b]$

One of the most important and widely studied classical Banach spaces is $C[a, b]$, the space of all continuous functions on the closed finite interval $[a, b]$, together with the norm

$$\|f\| = \max |f(x)|.$$

The celebrated Weierstrass approximation theorem asserts that the polynomials are *dense* in $C[a, b]$: if f is continuous on $[a, b]$, then for every positive number ε there is a polynomial P such that the inequality

$$|f(x) - P(x)| < \varepsilon$$

holds throughout the interval $[a, b]$.

For a given continuous function, a sequence of approximating polynomials can even be given explicitly. The most elegant representation is due to Bernstein. Let us suppose, for simplicity, that f is continuous on the interval $[0, 1]$. Then the nth *Bernstein polynomial* for f is

$$B_n(x) = \sum_{k=0}^{n} \binom{n}{k} f\left(\frac{k}{n}\right) x^k (1 - x)^{n-k}, \quad n = 1, 2, 3, \ldots.$$

As is well known,

$$f(x) = \lim_{n \to \infty} B_n(x)$$

uniformly on $[0, 1]$ (see Akhiezer [1956, p. 30]).

Since every polynomial can be uniformly approximated on a closed interval by a polynomial with *rational* coefficients, the preceding remarks show that the space $C[a, b]$ is separable; in fact, it has a basis.

Theorem 1 (Schauder). *The space $C[a, b]$ possesses a basis.*

Proof. We are going to construct a basis for $C[a, b]$ consisting of piecewise-linear functions f_n $(n = 0, 1, 2, \ldots)$. This means that to each function f in the space there will correspond a unique sequence of scalars $\{c_n\}$ such that

$$f(x) = \sum_{n=0}^{\infty} c_n f_n(x)$$

uniformly on $[a, b]$.

Let $\{x_0, x_1, x_2, \ldots\}$ be a countable dense subset of $[a, b]$ with $x_0 = a$ and $x_1 = b$. Set

$$f_0(x) = 1 \quad \text{and} \quad f_1(x) = \frac{x-a}{b-a}.$$

When $n \geqq 2$, the set of points $\{x_0, x_1, \ldots, x_{n-1}\}$ partitions $[a, b]$ into disjoint open intervals, one of which contains x_n; call it I. Define

$$f_n(x) = \begin{cases} 0 & \text{if} \quad x \notin I \\ 1 & \text{if} \quad x = x_n \\ \text{linear} & \text{elsewhere} \end{cases}$$

for $n = 2, 3, 4, \ldots$. The sequence $\{f_0, f_1, f_2, \ldots\}$ will be the required basis.

For each function f in $C[a, b]$ and each positive integer n, we denote by $L_n f$ the polygonal function that agrees with f at each of the points $x_0, x_1, \ldots, x_n$; we denote by $L_0 f$ the function whose constant value is $f(x_0)$. Since f is uniformly continuous on $[a, b]$, a simple continuity argument shows that

$$L_n f \to f \quad \text{uniformly on} \quad [a, b].$$

Therefore, we can write

$$f = L_0 f + \sum_{n=1}^{\infty} (L_n f - L_{n-1} f).$$

We are going to show that there are scalars $c_1, c_2, c_3, \ldots$ such that

$$L_n f - L_{n-1} f = c_n f_n \quad (n = 1, 2, 3, \ldots).$$

For this purpose, we shall define a sequence of functions $\{g_0, g_1, g_2, \ldots\}$ recursively by the equations

$$g_0 = f(x_0) f_0 \quad \text{and} \quad g_n = g_{n-1} + (f - g_{n-1})(x_n) f_n, \quad n = 1, 2, 3, \ldots.$$

The claim is that $g_n = L_n f$, whence

$$c_n = (f - L_{n-1} f)(x_n).$$

Since g_n is a polygonal function whose only possible corners are at the points $x_0, x_1, \ldots, x_n$, it is sufficient to show that g_n agrees with f at each of these points. This is trivial for $n = 0$, and we proceed by induction. Since $f_n(x_n) = 1$, it follows that $g_n(x_n) = f(x_n)$; if $i < n$, then $f_n(x_i) = 0$, and it follows from the definition of g_n, together with the induction hypothesis, that

$$g_n(x_i) = g_{n-1}(x_i) = f(x_i).$$

This establishes the claim.

Accordingly, every function $f \in C[a, b]$ has at least one representation of the form

$$f = \sum_{n=0}^{\infty} c_n f_n,$$

and we have only to show that this representation is unique. Suppose then that some function g has two different representations, say $\sum_{n=0}^{\infty} a_n f_n$ and $\sum_{n=0}^{\infty} b_n f_n$. If N is the smallest value of n for which $a_n \neq b_n$, then

$$\sum_{n=N}^{\infty} a_n f_n(x) = \sum_{n=N}^{\infty} b_n f_n(x)$$

for every x. Choose $x = x_N$. Since $f_n(x_N) = 0$ whenever $n > N$, it follows that $a_N = b_N$. But this contradicts the choice of N, and hence $a_n = b_n$ for every n. ■

PROBLEMS

1. Give a probabilistic interpretation of the Bernstein polynomials (*see* Feller [1966, Chap. VII]).
2. Prove that the space $C[a, b]$ is separable by showing that every continuous function on $[a, b]$ can be uniformly approximated by polynomials with *rational* coefficients.
3. Let f be a continuous function on $(-\infty, \infty)$. Prove that if there is a sequence of polynomials $\{P_1, P_2, P_3, \ldots\}$ such that $P_n \to f$ uniformly on $(-\infty, \infty)$, then f must itself be a polynomial.
4. Let f be a continuous function on $[a, b]$. Show that there is a sequence of polynomials $\{P_1, P_2, P_3, \ldots\}$ such that $f = \sum_{n=1}^{\infty} P_n$ and the series converges absolutely and uniformly on $[a, b]$.

3 ORTHONORMAL BASES IN HILBERT SPACE

In a separable Hilbert space†, a distinguished role is played by those Schauder bases that are *orthonormal*—the basis vectors are mutually perpendicular and each has unit length. An equivalent characterization of such bases is that they are *complete* orthonormal sequences. (Recall that a sequence

† All Hilbert spaces are assumed to be infinite-dimensional.

of vectors $\{f_1, f_2, f_3, \ldots\}$ in a Hilbert space is said to be complete if the zero vector alone is perpendicular to every f_n.) It follows readily from this characterization that every separable Hilbert space has an orthonormal basis.

The most important property of an orthonormal basis (as opposed to any other basis) is the simplicity of all basis expansions. If $\{e_1, e_2, e_3, \ldots\}$ is an orthonormal basis for a Hilbert space H, then for every element $f \in H$ we have the *Fourier expansion*

$$f = \sum_{n=1}^{\infty} (f, e_n) e_n.$$

The inner product (f, e_n) is called the nth *Fourier coefficient* of f (relative to $\{e_n\}$). When the Pythagorean formula is applied to this series, the result is **Parseval's identity:**

$$\|f\|^2 = \sum_{n=1}^{\infty} |(f, e_n)|^2.$$

The validity of Parseval's identity for every vector in the space is both necessary and sufficient for an orthonormal sequence to be a basis.

Since the linear transformation

$$f \to \{(f, e_n)\}$$

from H into l^2 preserves norms, it must also preserve inner products. Thus

$$(f, g) = \sum_{n=1}^{\infty} (f, e_n)\overline{(g, e_n)}$$

for every pair of vectors f and g; this is the *generalized* Parseval identity.

Even if an orthonormal sequence $\{e_n\}$ is incomplete, **Bessel's inequality** is always valid:

$$\sum_{n=1}^{\infty} |(f, e_n)|^2 \leqq \|f\|^2$$

whenever $f \in H$. This shows, in particular, that the Fourier coefficients of each element of H form a square-summable sequence. The **Riesz–Fischer theorem** shows, conversely, that every square-summable sequence is obtained in this way: if $\sum_{n=1}^{\infty} |c_n|^2 < \infty$, then there exists an element f in H for which

$$(f, e_n) = c_n, \quad n = 1, 2, 3, \ldots.$$

The proof is trivial: simply choose $f = \sum_{n=1}^{\infty} c_n e_n$. We conclude that if $\{e_n\}$ is a *complete* orthonormal sequence in H, then the correspondence $f \to \{(f, e_n)\}$ between H and l^2 is a Hilbert space isomorphism. It follows that from a geometric point of view, all separable Hilbert spaces are "indistinguishable", that is to say, isomorphic.

Example 1. In l^2 the "natural basis" is orthonormal.

Example 2. In $L^2[-\pi, \pi]$, with the inner product

$$(f, g) = \frac{1}{2\pi} \int_{-\pi}^{\pi} f(t)\overline{g(t)}\, dt,$$

the complex trigonometric system $\{e^{int}\}_{-\infty}^{\infty}$ constitutes an orthonormal basis. That the system is orthonormal is obvious; we prove that it is complete.

Theorem 2. *The trigonometric system is complete in* $L^2[-\pi, \pi]$.

Proof. The proof will establish even more. Suppose that

$$\int_{-\pi}^{\pi} f(t)e^{-int}\, dt = 0$$

for some *integrable* function f defined on $[-\pi, \pi]$ and $n = 0, \pm 1, \pm 2, \ldots$. It is to be shown that $f = 0$ a.e. Set

$$g(t) = \int_{-\pi}^{t} f(u)\, du$$

for $t \in [-\pi, \pi]$. Integration by parts shows that

$$\int_{-\pi}^{\pi} (g(t) - c)e^{-int}\, dt = 0$$

for every constant c and $n = \pm 1, \pm 2, \pm 3, \ldots$. Choose c so that this holds for $n = 0$ also, and put

$$F(t) = g(t) - c.$$

Then F is continuous on $[-\pi, \pi]$ and $F(\pi) = F(-\pi)$. Weierstrass's theorem on approximation by trigonometric polynomials guarantees that for each $\varepsilon > 0$ there is a finite trigonometric sum

$$T(t) = \sum_{k=-n}^{n} c_k e^{ikt}$$

such that

$$|F(t) - T(t)| < \varepsilon \quad \text{whenever} \quad |t| \leqq \pi.$$

It follows that

$$\|F\|^2 = \frac{1}{2\pi} \int_{-\pi}^{\pi} |F(t)|^2\, dt = \frac{1}{2\pi} \int_{-\pi}^{\pi} \overline{F(t)}(F(t) - T(t))\, dt$$

$$\leqq \frac{\varepsilon}{2\pi} \int_{-\pi}^{\pi} |F(t)|\, dt \leqq \varepsilon \|F\|,$$

so that

$$\|F\| \leqq \varepsilon.$$

Since ε was arbitrary, $F = 0$, so that $g = c$ and $f = 0$ a.e. ∎

Consequently, every function f in $L^2[-\pi, \pi]$ has a unique Fourier series expansion

$$f(t) = \sum_{-\infty}^{\infty} \hat{f}(n) e^{int}$$

(*in the mean*†). Here $\hat{f}(n)$ denotes the nth Fourier coefficient of f relative to $\{e^{int}\}$, i.e.,

$$\hat{f}(n) = \frac{1}{2\pi} \int_{-\pi}^{\pi} f(t) e^{-int}\, dt \quad (n = 0, \pm 1, \pm 2, \ldots).$$

By Parseval's formula,

$$\frac{1}{2\pi} \int_{-\pi}^{\pi} |f(t)|^2\, dt = \sum_{-\infty}^{\infty} |\hat{f}(n)|^2.$$

The mapping $f \to \{\hat{f}(n)\}$ is a Hilbert space isomorphism between $L^2[-\pi, \pi]$ and l^2.

There is a simple but useful extension of Parseval's identity that is worth mentioning. If $f \in L^2[-\pi, \pi]$, let $\hat{f}$ be the **Fourier transform** of f:

$$\hat{f}(x) = \frac{1}{2\pi} \int_{-\pi}^{\pi} f(t) e^{-ixt}\, dt \quad (-\infty < x < \infty).$$

Proposition 1. *For every function $f \in L^2[-\pi, \pi]$ and every real number A,*

$$\sum_{-\infty}^{\infty} |\hat{f}(n + A)|^2 = \|f\|^2.$$

Proof. Put $g(t) = f(t) e^{-iAt}$. Then

$$\hat{f}(n + A) = \hat{g}(n)$$

for every integer n. Since A is real, $\|f\| = \|g\|$, and the result follows from Parseval's identity applied to g. ∎

As an illustration, let us choose f to be the constant function 1. A simple calculation shows that

$$\hat{f}(x) = \frac{\sin \pi x}{\pi x}$$

† Pointwise convergence is of course much harder. A deep result of Carleson [1966] says that the Fourier series of an L^2 function converges (to the function) pointwise almost everywhere.

for all real x. Setting $A = t/\pi$, where t is real and not an integral multiple of π, we obtain the important identity

$$\frac{1}{\sin^2 t} = \sum_{-\infty}^{\infty} \frac{1}{(n\pi + t)^2}.$$

Example 3. The space $\boldsymbol{H}^2$ (named after Hardy) consists of all functions f analytic in the open unit disk (in the complex plane) whose Taylor coefficients are square-summable, i.e.,

$$f(z) = \sum_{n=0}^{\infty} c_n z^n, \quad \text{with} \quad \sum_{n=0}^{\infty} |c_n|^2 < \infty.$$

The inner product of two functions $f(z) = \sum_{n=0}^{\infty} a_n z^n$ and $g(z) = \sum_{n=0}^{\infty} b_n z^n$ in $\boldsymbol{H}^2$ is, by definition,

$$(f, g) = \sum_{n=0}^{\infty} a_n \overline{b}_n.$$

It is clear that $\boldsymbol{H}^2$ can be identified with the (closed) subspace of $L^2[-\pi, \pi]$ spanned by the functions e^{int} with $n \geqq 0$.

Let $e_n(z) = z^n$ for $|z| < 1$ $(n = 0, 1, 2, \ldots)$; then the e_n's form an orthonormal basis for $\boldsymbol{H}^2$. The natural mapping

$$(c_0, c_1, c_2, \ldots) \to \sum_{n=0}^{\infty} c_n z^n$$

between l^2 and $\boldsymbol{H}^2$ is a Hilbert space isomorphism.

Example 4. The space $\boldsymbol{A}^2$ consists of all functions f that are analytic in the open unit disk and have finite area norm

$$\|f\| = \left(\iint_{|z|<1} |f(z)|^2 \, dx \, dy \right)^{1/2}.$$

Under pointwise addition and scalar multiplication, $\boldsymbol{A}^2$ is a vector space; when endowed with the inner product

$$(f, g) = \iint_{|z|<1} f(z)\overline{g(z)} \, dx \, dy,$$

it becomes a Hilbert space (see Problem 16).

Assertion: If

$$e_n(z) = \sqrt{\frac{n+1}{\pi}} z^n \quad \text{for} \quad |z| < 1 \quad (n = 0, 1, 2, \ldots),$$

then the e_n's form an orthonormal basis for $\boldsymbol{A}^2$. The orthonormal part is simple: setting $z = re^{i\theta}$, we have

$$(e_n, e_m) = \frac{1}{\pi}\sqrt{n+1}\sqrt{m+1}\int_0^{2\pi}\int_0^1 e^{i(n-m)\theta} r^{n+m} r\,dr\,d\theta$$
$$= \frac{2\sqrt{n+1}\sqrt{m+1}}{n+m+2}\delta_{nm} = \delta_{nm}.$$

As for completeness, it must be shown that if $f \in \boldsymbol{A}^2$ and $(f, e_n) = 0$ for $n \geqq 0$, then $f = 0$. This we accomplish by proving that if the Taylor series of f is

$$f(z) = \sum_{n=0}^{\infty} c_n z^n,$$

then

$$c_n = \sqrt{\frac{n+1}{\pi}}(f, e_n) \quad (n = 0, 1, 2, \ldots).$$

Argue as follows:

$$f(z)\bar{z}^m = \sum_{n=0}^{\infty} c_n z^n \bar{z}^m.$$

Since the series is uniformly convergent in each disk $\{z\colon |z| \leqq r\}$, $0 < r < 1$, term-by-term integration is permissible and yields

$$\iint_{|z|\leqq r} f(z)\bar{z}^m\,dx\,dy = 2\pi \sum_{n=0}^{\infty} c_n \delta_{nm} \frac{r^{n+m+2}}{n+m+2} = \pi c_m \frac{r^{2m+2}}{m+1}.$$

Since $|f(z)\bar{z}^m|$ is integrable over $\{z\colon |z| < 1\}$,

$$(f, z^m) = \lim_{r\to 1} \iint_{|z|\leqq r} f(z)\bar{z}^m\,dx\,dy = \frac{\pi c_m}{m+1},$$

and the result follows.

Let us remark that we now know that the Taylor series of a function f in $\boldsymbol{A}^2$ converges to f in the topology of $\boldsymbol{A}^2$. This could not have been deduced solely from the fact that the Taylor series of f is known to converge (to f) uniformly on each closed subset of the disk (see Problem 17).

If $f \in \boldsymbol{A}^2$, with Taylor series $f(z) = \sum_{n=0}^{\infty} c_n z^n$, then Parseval's formula shows that

$$\|f\|^2 = \pi \sum_{n=0}^{\infty} \frac{|c_n|^2}{n+1}.$$

This provides an alternative description of $\boldsymbol{A}^2$ as the class of all functions $f(z) = \sum_{n=0}^{\infty} c_n z^n$ analytic in the open unit disk for which

$$\sum_{n=0}^{\infty} \frac{|c_n|^2}{n+1} < \infty.$$

The inner product of two such functions $\sum_{n=0}^{\infty} a_n z^n$ and $\sum_{n=0}^{\infty} b_n z^n$ is in this case defined to be

$$\pi \sum_{n=0}^{\infty} \frac{a_n \bar{b}_n}{n+1},$$

in agreement with the original definition.

PROBLEMS

1. Prove that every separable Hilbert space contains a complete orthonormal sequence.
2. Give an example of a nonorthogonal basis for a separable Hilbert space.
3. Let $\{e_1, \ldots, e_n\}$ be an orthonormal subset of a Hilbert space H and f an arbitrary element of H. Show that the minimum value of $\left\| f - \sum_{i=1}^{n} c_i e_i \right\|$ is attained when and only when $c_i = (f, e_i)$ for $i = 1, \ldots, n$.
4. Let $\{e_1, e_2, e_3, \ldots\}$ be a complete orthonormal sequence in a Hilbert space H and suppose that $f_1, f_2, f_3, \ldots$ are arbitrary vectors in H such that

$$\sum_{n=1}^{\infty} \|e_n - f_n\|^2 < 1.$$

   Show that the sequence $\{f_1, f_2, f_3, \ldots\}$ is also complete in H.
5. Prove that an orthonormal basis for a separable Hilbert space is *unconditional.* (For the definition of an unconditional basis, see Problem 9 of Section 1.)
6. Let X be a Banach space and x and y elements of X. We say that x is orthogonal to y, and we write $x \perp y$, if

$$\|x\| \leqq \|x + \lambda y\| \quad \text{for all scalars } \lambda.$$

   (a) Show by an example that $x \perp y$ need not imply that $y \perp x$.
   (b) Show that the relations $x \perp y$ and $x \perp z$ need not imply that $x \perp (y + z)$.
   (c) Show that if X is a Hilbert space, then $x \perp y$ if and only if $(x, y) = 0$.
   (d) Is every complete orthogonal sequence in X a basis?
7. Let H be a nonseparable Hilbert space.
   (a) Using the Hausdorff Maximal Principle, show that H contains a complete orthonormal subset.
   (b) Let $\{e_\alpha : \alpha \in A\}$ be an orthonormal subset of H and f an arbitrary element of H. Show that $(f, e_\alpha) = 0$ for all but countably many values of α.
   (c) Show that every complete orthonormal subset of H has the same number of elements. This number is called the (orthogonal or Hilbert space) **dimension** of H.
   (d) Show that two Hilbert spaces are isomorphic if and only if they have the same dimension.

(e) Show that for each cardinal number α, there is a Hilbert space whose dimension is α.

8. Let X be the vector space of all finite linear combinations of functions of the form $e^{i\lambda t}(-\infty < t < \infty)$, where the parameter λ is *real*. An inner product in X is defined by

$$(f, g) = \lim_{T\to\infty} \frac{1}{2T}\int_{-T}^{T} f(t)\overline{g(t)}\,dt.$$

When X is closed by means of the metric generated by this inner product, we obtain a certain Hilbert space B^2 (B is for Besicovitch).

(a) Show that the continuum of elements $e^{i\lambda t}$ forms a complete orthonormal subset of B^2.

(b) Show that B^2 contains the important class of *Bohr* **almost periodic** functions. These are obtained by adding to X the limits of sequences of functions in X that are uniformly convergent on the entire real line.

9. (a) Show that each of the systems $\{\sqrt{2}\sin nt : n = 1, 2, 3, \ldots\}$ and $\{1, \sqrt{2}\cos nt : n = 1, 2, 3, \ldots\}$ forms an orthonormal basis for $L^2[0, \pi]$.

(b) Show that the system $\{1, \cos nt, \sin(n - \frac{1}{2})t : n = 1, 2, 3, \ldots\}$ forms an orthonormal basis for $L^2[-\pi, \pi]$.

10. Prove that the sequence of functions $\{1, x, x^2, \ldots\}$ is complete in $L^2[a, b]$ for every finite interval $[a, b]$.

11. Show that an orthonormal sequence $\{e_1, e_2, e_3, \ldots\}$ in a separable Hilbert space H is complete if and only if Parseval's identity

$$\sum_{n=1}^{\infty} |(f, e_n)|^2 = \|f\|^2$$

holds for every vector f belonging to a *complete* subset of H.

12. **(Vitali)** An orthonormal sequence $\{e_n\}$ in $L^2[a, b]$ is complete if and only if

$$\sum_{n=1}^{\infty} \left|\int_a^x e_n(t)\,dt\right|^2 = x - a$$

for every x in $[a, b]$.

13. **(Dalzell)** An orthonormal sequence $\{e_n\}$ in $L^2[a, b]$ is complete if and only if

$$\sum_{n=1}^{\infty} \int_a^b \left|\int_a^x e_n(t)\,dt\right|^2 dx = \frac{(b-a)^2}{2}.$$

14. Show that when the *Gram–Schmidt* process is applied to the functions $1, x, x^2, \ldots$ in $L^2[-1, 1]$, the resulting (complete) orthonormal sequence is $\sqrt{n+\frac{1}{2}}P_n(x)$, where

$$P_n(x) = \frac{1}{2^n n!}\frac{d^n}{dx^n}(x^2-1)^n \quad (n = 0, 1, 2, \ldots)$$

is the nth **Legendre polynomial**.

15. Show that the space $\boldsymbol{H}^2$ can be described alternatively as the class of all functions f analytic in the open unit disk for which the integral means

$$\left(\frac{1}{2\pi}\int_0^{2\pi} |f(re^{i\theta})|^2\, d\theta\right)^{1/2}$$

remain bounded as $r \to 1$.

16. (a) Show that if $f \in \boldsymbol{A}^2$, then

$$|f(z)| \leqq \frac{\|f\|}{\sqrt{\pi}(1-|z|)} \quad \text{whenever} \quad |z| < 1.$$

(*Hint*: The value of an analytic function at the center of a disk is equal to its average value over the entire disk.)

(b) Prove that $\boldsymbol{A}^2$ is a Hilbert space.

17. Show that if $f_n \to f$ in $\boldsymbol{A}^2$, then $f_n \to f$ uniformly on every closed subset of $|z| < 1$. Show by an example that the converse is false.

4 THE REPRODUCING KERNEL

Most of the important examples of Hilbert spaces are function spaces. The special properties of the functions considered, such as analyticity, enrich the structure of the space, which in return supplies added information about the functions.

Let H be a Hilbert space whose elements are real or complex-valued functions defined on a set S. We shall call H a **functional Hilbert space** if for every element $x \in S$ the "point-evaluation" functional

$$f \to f(x)$$

on H is bounded. This means that there is a constant M_x such that for all $f \in H$ we have $|f(x)| \leqq M_x\|f\|$. By the Riesz representation theorem, every bounded linear functional on H arises from an inner product, and so if $x \in S$, there is an element $K_x \in H$ such that

$$f(x) = (f, K_x) \quad \text{for every } f.$$

The function K on $S \times S$, defined by

$$K(x, y) = (K_y, K_x) = K_y(x),$$

is called the **kernel function** or the **reproducing kernel** of H.

Example 1. The sequence space l^2 provides a trivial example of a functional Hilbert space. Here S is the set of natural numbers, and K is given by

$$K(m, n) = (e_n, e_m) = \delta_{mn}$$

($\{e_n\}$ is the "natural basis" for l^2).

Example 2. Consider the space $\boldsymbol{H}^2$. For $|\beta| < 1$ the function g defined by

$$g(z) = \sum_{n=0}^{\infty} \overline{\beta}^n z^n$$

belongs to $\boldsymbol{H}^2$ and

$$f(\beta) = (f, g) \quad \text{for every } f \in \boldsymbol{H}^2.$$

This shows that "evaluation at β" is a bounded linear functional on $\boldsymbol{H}^2$. It also shows that $g = K_\beta$, so that the reproducing kernel of $\boldsymbol{H}^2$ is given by

$$K(z, w) = \sum_{n=0}^{\infty} z^n \overline{w}^n = \frac{1}{1 - z\overline{w}};$$

K is called the *Szegö kernel.*

The next proposition shows that the reproducing kernel of a (separable) functional Hilbert space can always be described explicitly in terms of an orthonormal basis for the space.

Proposition 2. *If $\{e_1, e_2, e_3, \ldots\}$ is an orthonormal basis for a functional Hilbert space H and if K is the kernel function of H, then*

$$K(x, y) = \sum_{n=1}^{\infty} e_n(x)\overline{e_n(y)}.$$

Proof. Let x be a fixed but arbitrary element of S. If K_x is expanded in a Fourier series relative to $\{e_n\}$, then

$$K_x = \sum_{n=1}^{\infty} (K_x, e_n)e_n = \sum_{n=1}^{\infty} \overline{e_n(x)} e_n.$$

The result follows at once from Parseval's identity (generalized) since

$$K(x, y) = (K_y, K_x) = \sum_{n=1}^{\infty} e_n(x)\overline{e_n(y)}. \quad \blacksquare$$

Example 3. Consider the space $\boldsymbol{A}^2$. That $\boldsymbol{A}^2$ is a *functional Hilbert space* is an easy consequence of the following mean-value formula for analytic functions. If f is analytic in the closed disk $D_r = \{z\colon |z - z_0| \leqq r\}$, then

$$f(z_0) = \frac{1}{\pi r^2} \iint_{D_r} f(z)\, dx\, dy.$$

To prove the formula, observe that for every $\rho \leqq r$,

$$f(z_0) = \frac{1}{2\pi}\int_0^{2\pi} f(z_0 + \rho e^{i\theta})\,d\theta,$$

by Cauchy's integral formula. Multiplying both sides by ρ and then integrating (with respect to ρ) from 0 to r, we obtain

$$f(z_0) = \frac{2}{r^2}\int_0^r f(z_0)\rho\,d\rho = \frac{1}{\pi r^2}\int_0^{2\pi}\int_0^r f(z)\rho\,d\rho\,d\theta,$$

which is the desired result.

Suppose now that $f \in \boldsymbol{A}^2$ and $|z_0| < 1$. Choose r small enough so that D_r is completely contained within the open unit disk. When the Cauchy–Schwarz inequality is applied to this mean-value formula, the result is

$$|f(z_0)| \leqq \frac{1}{\pi r^2}\iint_{D_r} |f(z)|\,dx\,dy \leqq \frac{1}{\sqrt{\pi}r}\|f\|.$$

This shows that "evaluation at z_0" is a bounded linear functional on $\boldsymbol{A}^2$.

To determine the reproducing kernel of $\boldsymbol{A}^2$, apply Proposition 2. The functions e_n defined by $e_n(z) = \sqrt{(n+1)/\pi}z^n$ for $|z| < 1$ and $n = 0, 1, 2, \ldots$ form an orthonormal basis for $\boldsymbol{A}^2$ (see Section 3), and hence

$$K(z, w) = \frac{1}{\pi}\sum_{n=0}^{\infty}(n+1)z^n\overline{w}^n = \frac{1}{\pi(1 - z\overline{w})^2},$$

for $|z| < 1$, $|w| < 1$. The kernel function K of $\boldsymbol{A}^2$ is called the *Bergman kernel.*

The following integral representation for a function f in $\boldsymbol{A}^2$ is now immediate:

$$\begin{aligned} f(w) = (f(z), K(z, w)) &= \iint_{|z|<1} f(z)\overline{K(z, w)}\,dx\,dy \\ &= \frac{1}{\pi}\iint_{|z|<1} f(z)\frac{1}{(1 - \overline{z}w)^2}\,dx\,dy. \end{aligned}$$

PROBLEMS

1. Is the space $L^2[0, 1]$ a functional Hilbert space?

2. Let $\{e_n\}$ be an orthonormal sequence in a functional Hilbert space H, with reproducing kernel K. Show that $\{e_n\}$ is complete if and only if

$$K(x, x) = \sum_{n=1}^{\infty} |e_n(x)|^2$$

for every x.

3. A function K defined on $S \times S$ is called a **positive matrix** if for each positive integer n and each choice of points $t_1, \ldots, t_n$ from S the quadratic form

$$\sum_{j=1}^{n} \sum_{i=1}^{n} K(t_i, t_j)\xi_i \overline{\xi}_j$$

is positive definite.

(a) Show that the reproducing kernel of a functional Hilbert space is a positive matrix.

(b) Show that if K is a positive matrix, then there is a functional Hilbert space whose reproducing kernel is K.

4. Let $\boldsymbol{P}$ be the Hilbert space of all entire functions of the form

$$f(z) = \frac{1}{2\pi} \int_{-\pi}^{\pi} \varphi(t) e^{-izt}\, dt,$$

where $\varphi \in L^2[-\pi, \pi]$. The inner product of two functions f and g in $\boldsymbol{P}$ is defined to be

$$(f, g) = \int_{-\infty}^{\infty} f(x)\overline{g(x)}\, dx.$$

Show that $\boldsymbol{P}$ is a functional Hilbert space and find its reproducing kernel. ($\boldsymbol{P}$ is the famous *Paley–Wiener space*; it will play an important role in the theory of nonharmonic Fourier series.)

5 COMPLETE SEQUENCES

A basis for a Banach space has the important property that every vector in the space can be approximated arbitrarily closely by finite combinations of the basis elements. Of course, a basis is much more. In many cases, however, it is sufficient to know that a sequence of vectors, although not necessarily a basis, does nevertheless have this approximation property. Such sequences will be called *complete*.

Definition. *A sequence of vectors $\{x_1, x_2, x_3, \ldots\}$ in a normed vector space X is said to be* ***complete***† *if its linear span is dense in X, that is, if for each vector x and each $\varepsilon > 0$ there is a finite linear combination $c_1x_1 + \cdots + c_nx_n$ such that*

$$\|x - (c_1x_1 + \cdots + c_nx_n)\| < \varepsilon.$$

† The terminology of the subject is not uniform—the terms "closed", "total", and "fundamental" are also used.

It is a direct and important consequence of the Hahn–Banach theorem that the completeness of a sequence of vectors $\{x_n\}$ in X is equivalent to the following condition: if $\mu \in X^*$ (the *topological dual* of X) and if $\mu(x_n) = 0$ $(n = 1, 2, 3, \ldots)$, then $\mu = 0$. When X is a Hilbert space, the Riesz representation theorem shows that all is well—the former definition (see Section 3) agrees with the present one.

Example 1. If $(X, \mathscr{B}, \mu)$ is a σ-finite measure space and $1 \leqq p < \infty$, then the Riesz representation theorem shows that the dual of $L^p(\mu)$ can be identified with $L^q(\mu)$, where $1/p + 1/q = 1$. It follows that a sequence $\{f_n\}$ of functions in $L^P(\mu)$ will be complete provided the relations

$$\int_X f_n g \, d\mu = 0 \quad (n = 1, 2, 3, \ldots),$$

with $g \in L^q(\mu)$, imply that $g = 0$ a.e.

If μ is a finite measure, then $L^P(\mu) \subset L^1(\mu)$ so that if the relations above imply that $g = 0$ a.e. whenever $g \in L^1(\mu)$, then the sequence $\{f_n\}$ will be complete in $L^P(\mu)$ for every $p \geqq 1$.

Example 2. The trigonometric system $\{e^{int}\}_{-\infty}^{\infty}$ is complete in $L^P[-\pi, \pi]$ for $1 \leqq p < \infty$. Indeed, the proof of Theorem 2 showed that if $f \in L^1[-\pi, \pi]$ and

$$\int_{-\pi}^{\pi} f(t)e^{-int} \, dt = 0 \quad (n = 0, \pm 1, \pm 2, \ldots),$$

then $f = 0$ a.e.

Example 3. In $C[a, b]$ the sequence of powers $\{1, t, t^2, \ldots\}$ is complete by virtue of the Weierstrass theorem on polynomial approximation: the polynomials are dense in $C[a, b]$. It follows that if g is an element of $C[a, b]$ for which all of the "moments"

$$\int_a^b t^n g(t) \, dt \quad (n = 0, 1, 2, \ldots)$$

are zero, then g must be identically zero. Indeed, if we write

$$\mu(f) = \int_a^b f(t) g(t) \, dt,$$

then μ is a bounded linear functional on $C[a, b]$ and $\mu(t^n) = 0$ for every n. Since the powers of t are complete, $\mu = 0$, and hence g must be identically zero.

Example 4. The sequence of powers $\{1, t, t^2, \ldots\}$ is complete in $L^p[a, b]$ whenever $[a, b]$ is an arbitrary finite interval and $1 \leqq p < \infty$. This follows readily from Examples 1 and 3. For suppose $f \in L^1[a, b]$ and

$$\int_a^b t^n f(t)\, dt = 0 \quad (n = 0, 1, 2, \ldots).$$

If we define

$$F(t) = \int_a^t f(u)\, du, \quad a \leqq t \leqq b,$$

then $F \in C[a, b]$ and $F(a) = F(b) = 0$. Integration by parts shows that

$$0 = \int_a^b t^n f(t)\, dt = -n \int_a^b t^{n-1} F(t)\, dt$$

for $n = 1, 2, 3, \ldots$. Conclusion: $F = 0$ everywhere so that $f = 0$ almost everywhere.

PROBLEMS

1. Show that if $\{e_n\}$ is a complete orthonormal sequence in a Hilbert space, then $\{e_n - e_{n+1}\}$ is also complete. Is the result true if $\{e_n\}$ is merely assumed to be complete?
2. Let $\{x_n\}$ be a complete sequence in a normed vector space X. Show that if $\{y_n\}$ is complete in $\{x_n\}$ (that is, if each x_n can be approximated arbitrarily closely by linear combinations of the y_n), then $\{y_n\}$ is complete in X.
3. Show that the sequence $\{1, x^2, x^4, \ldots\}$ is complete in $C[0, 1]$. (Give an "elementary" proof, that is, one that uses neither the Müntz–Szász theorem nor the Stone–Weierstrass theorem.)
4. Show that if $\{f_1, f_2, f_3, \ldots\}$ is a sequence in $C[0, 1]$ that is complete in $L^2[0, 1]$, then $\{1, f_1, f_2, \ldots\}$ is complete in $C[0, 1]$.
5. Show that the trigonometric system $\{e^{int}\}_{-\infty}^{\infty}$ is not complete in $L^2[-A, A]$ if $A > \pi$.
6. **(Boas-Pollard)** Let $\{f_0, f_1, f_2, \ldots\}$ be a complete orthonormal sequence in $L^2[a, b]$. Show that there exists a bounded measurable function g such that the sequence $\{gf_1, gf_2, gf_3, \ldots\}$ is complete in $L^2[a, b]$. (*Hint*: Any bounded measurable function g that is never zero and is such that $f_0/g \notin L^2[a, b]$ will do.)
7. Show that the sequence $\left\{\frac{1}{x+1}, \frac{1}{x+2}, \frac{1}{x+3}, \ldots\right\}$ is complete in $L^2[0, 1]$.
8. Let D be a bounded *domain* (an open connected set) in the complex plane and $A(D)$ the space of functions f that are analytic in D and continuous in the closure of D. Define $\|f\| = \max\{|f(z)|: z \in \overline{D}\}$.
   (a) Show that if D is the open unit disk, $D = \{z: |z| < 1\}$, then the sequence $\{1, z, z^2, \ldots\}$ is complete in $A(D)$.

(b) Show that if D is multiply-connected, then the sequence $\{1, z, z^2, \ldots\}$ is not complete in $A(D)$.

9. Let T be the unit circle in the complex plane, $T = \{z: |z| = 1\}$, and $C(T)$ the space of all continuous complex-valued functions f on T. Define $\|f\| = \max\{|f(z)|: z \in T\}$. Show that the polynomials in z are not dense in $C(T)$. (Give an example of a continuous function on T that is not the uniform limit of polynomials.)

6 THE COEFFICIENT FUNCTIONALS

If $\{x_1, x_2, x_3, \ldots\}$ is a basis for a Banach space X, then every vector x in the space has a unique series expansion of the form

$$x = \sum_{n=1}^{\infty} c_n x_n.$$

It is clear that each coefficient c_n is a linear function of x. If we denote this linear function by f_n, then $c_n = f_n(x)$, and we may write

$$x = \sum_{n=1}^{\infty} f_n(x) x_n.$$

The functionals f_n $(n = 1, 2, 3, \ldots)$ thus defined are called the **coefficient functionals** associated with the basis $\{x_n\}$. They play an essential role in many parts of Banach space theory; perhaps the most important statement about them is that they are continuous.

Theorem 3. *If $\{x_n\}$ is a basis for a Banach space X and if $\{f_n\}$ is the associated sequence of coefficient functionals, then each $f_n \in X^*$. Moreover, there exists a constant M such that*

$$1 \leqq \|x_n\| \cdot \|f_n\| \leqq M \quad (n = 1, 2, 3, \ldots).$$

Proof. Introduce the vector space Y consisting of those sequences of scalars $\{c_n\}$ for which the series $\sum_{n=1}^{\infty} c_n x_n$ is convergent in X. If $\{c_n\} \in Y$, then the number

$$\|\{c_n\}\| = \sup_n \left\| \sum_{i=1}^{n} c_i x_i \right\|$$

satisfies all the properties of a norm. We begin by showing that Y, endowed with this norm, is a Banach space isomorphic to X. Verification that Y is a Banach space is typical of standard completeness arguments, and since it poses no special difficulties, it is left as an exercise for the reader. To see that

X and Y are isomorphic, argue as follows. The mapping $T: Y \to X$ defined by

$$\{c_n\} \to \sum_{n=1}^{\infty} c_n x_n$$

is obviously linear; since $\{x_n\}$ is a basis, it is also one-to-one and onto. Since

$$\left\| \sum_{n=1}^{\infty} c_n x_n \right\| \leqq \sup_n \left\| \sum_{i=1}^{n} c_i x_i \right\|,$$

it follows that T is continuous, and the open mapping theorem then guarantees that T^{-1} is also continuous. This proves that X and Y are isomorphic.

Suppose now that $x = \sum_{n=1}^{\infty} c_n x_n$ is a fixed but arbitrary element of X. Then for every n,

$$|f_n(x)| = |c_n| = \frac{\|c_n x_n\|}{\|x_n\|} \leqq \frac{\left\| \sum_{i=1}^{n} c_i x_i \right\| + \left\| \sum_{i=1}^{n-1} c_i x_i \right\|}{\|x_n\|}$$

$$\leqq \frac{2 \sup_n \left\| \sum_{i=1}^{n} c_i x_i \right\|}{\|x_n\|} = \frac{2\|T^{-1}x\|}{\|x_n\|} \leqq \frac{2\|T^{-1}\| \cdot \|x\|}{\|x_n\|}.$$

This proves that each f_n is continuous and that $\|f_n\| \leqq 2\|T^{-1}\|/\|x_n\|$. Choosing $M = 2\|T^{-1}\|$, we have

$$\|x_n\| \cdot \|f_n\| \leqq M$$

for every n. The remaining inequality is trivial:

$$1 = f_n(x_n) \leqq \|f_n\| \cdot \|x_n\|. \blacksquare$$

Definition. *Let $\{x_n\}$ be a basis for a Banach space X and let $\{f_n\}$ be the associated sequence of coefficient functionals. For $n = 1, 2, 3, \ldots$ the nth* ***partial sum operator*** *S_n is the linear operator on X defined by*

$$S_n(x) = \sum_{i=1}^{n} f_i(x) x_i.$$

Theorem 3 implies that each partial sum operator is bounded, and a closer examination of the proof reveals further that

$$1 \leqq \sup_n \|S_n\| < \infty.$$

As an application of this fact, we establish a simple and useful criterion for determining when a complete sequence is a basis.

Theorem 4. *A complete sequence $\{x_n\}$ of nonzero vectors in a Banach space X is a basis for X if and only if there exists a constant M such that*

$$\left\| \sum_{i=1}^{n} c_i x_i \right\| \leqq M \left\| \sum_{i=1}^{m} c_i x_i \right\|$$

whenever $c_1, \ldots, c_m$ are arbitrary scalars and $n \leqq m$.

Proof. The necessity is easy: we need only choose $M = \sup_n \|S_n\|$, where S_n is the nth partial sum operator associated with $\{x_n\}$. Reason: If $m \geqq n$, then

$$\sum_{i=1}^{n} c_i x_i = S_n \left(\sum_{i=1}^{m} c_i x_i \right);$$

now take the norm of both sides.

For the sufficiency, start with a fixed vector x and select scalars $\{c_{in}\}_{i=1}^{n}$, $n = 1, 2, 3, \ldots$, such that

$$x = \lim_{n \to \infty} \sum_{i=1}^{n} c_{in} x_i .$$

(This is possible because $\{x_n\}$ is complete.) For notational convenience, put $c_{in} = 0$ for $i > n$. For fixed k and $m > n > k$ we have by hypothesis

$$\begin{aligned} \|(c_{kn} - c_{km}) x_k\| &\leqq M \left\| \sum_{i=1}^{n} (c_{in} - c_{im}) x_i \right\| \leqq M^2 \left\| \sum_{i=1}^{m} (c_{in} - c_{im}) x_i \right\| \\ &= M^2 \left\| \sum_{i=1}^{n} c_{in} x_i - \sum_{i=1}^{m} c_{im} x_i \right\| . \end{aligned}$$

As $m, n \to \infty$, this last expression tends to zero. Since $x_k \neq 0$, it follows that for each k the sequence $\{c_{kn}\}_{n=1}^{\infty}$ is a Cauchy sequence, and therefore the limit

$$c_k = \lim_{n \to \infty} c_{kn}$$

exists. It is now a routine matter to show that

$$x = \sum_{k=1}^{\infty} c_k x_k .$$

The details are left to the reader. We complete the proof by showing that this representation is unique.

It suffices to show that if $\sum_{k=1}^{\infty} c_k x_k = 0$, then $c_k = 0$ for every k. For fixed k and $n \geqq k$ we have once again by hypothesis

$$\|c_k x_k\| \leqq M \left\| \sum_{i=1}^{n} c_i x_i \right\| .$$

Letting $n \to \infty$, we find $\|c_k x_k\| = 0$, and hence $c_k = 0$ for every k. ∎

PROBLEMS

1. Show that the normed vector space Y defined in the proof of Theorem 3 is a Banach space.
2. Let $\{x_n\}$ denote Schauder's basis for $C[a, b]$ and let $\{f_n\}$ be the associated sequence of coefficient functionals. Compute $\|f_n\|$.
3. Use the criterion of Theorem 4 to show that Schauder's system is in fact a basis for $C[a, b]$.
4. Let $\{x_n\}$ be a basis for a Banach space X and $\{S_n\}$ the corresponding sequence of partial sum operators. Show that $1 \leqq \sup_n \|S_n\| < \infty$.
5. **(Karlin)** A basis $\{x_n\}$ for a Banach space X is said to be **normal** if $\|x_n\| = \|f_n\| = 1$ for every n. Prove that in a Hilbert space every normal basis is orthonormal. (*Hint*: Show that if $\{x_n\}$ is normal, then $x_n \perp x_m$ whenever $n \neq m$ (see Problem 6, Section 3).)
6. A sequence $\{x_n\}$ of *nonzero* vectors in a Banach space X is said to be an *orthogonal system* if for each n $(n = 1, 2, 3, \ldots)$ the linear span of $\{x_1, \ldots, x_n\}$ is orthogonal to x_{n+1} (see Problem 6, Section 3).
   (a) Show that an orthogonal system is linearly independent.
   (b) Show that a complete orthogonal system is a basis.
7. A Banach space X is said to have the **approximation property** (*in the sense of Grothendieck*) if the identity operator on X can be approximated, uniformly on every compact subset of X, by operators of finite rank.
   (a) Prove that every Banach space with a basis has the approximation property.
   (b) Let X be a Banach space with a basis and let Y be an arbitrary Banach space. Prove that every compact linear transformation from X into Y is the limit (with respect to the *norm topology*) of operators of finite rank.
8. Prove that the *disk algebra* A has the approximation property. (See Problem 7. In the notation of Problem 8, Section 5, $A = A(D)$, with $D = \{z: |z| < 1\}$.)

7 DUALITY

Suppose that $\{x_n\}$ is a basis for a Banach space X and that $\{f_n\}$ is its associated sequence of coefficient functionals. What can we say about the sequence $\{f_n\}$? Surely not that it is a basis for X^*. For if X^* is nonseparable, then it contains no basis at all (example: $X = l^1$). Therefore, unless $\{f_n\}$ is complete, the most we can hope for is that it be a basis for its closed linear span. The hope is justified.

The closed linear span of a sequence $\{x_n\}$ of elements from a normed vector space X will be denoted by $[x_n]$.

Theorem 5. *Let $\{x_n\}$ be a basis for a Banach space X and let $\{f_n\}$ be the associated sequence of coefficient functionals. Then $\{f_n\}$ is a basis for $[f_n]$*

and the expansion

$$f = \sum_{n=1}^{\infty} f(x_n) f_n$$

is valid for every f in $[f_n]$.

Proof. The proof is based on the fact that the adjoint of the nth partial sum operator S_n is given by the formula

$$S_n^* f = \sum_{i=1}^{n} f(x_i) f_i$$

for every $f \in X^*$. The formula is valid because

$$(S_n^* f)(x) = f(S_n x) = f\left(\sum_{i=1}^{n} f_i(x) x_i\right) = \left(\sum_{i=1}^{n} f(x_i) f_i\right)(x)$$

for every $x \in X$. It is to be shown that $S_n^* f \to f$ for every $f \in [f_n]$.

Let us first assume that f is a finite linear combination of the f_i, say $f = \sum_{i=1}^{m} c_i f_i$. Then for every $n \geqq m$

$$S_n^* f = \sum_{i=1}^{n} f(x_i) f_i = \sum_{i=1}^{m} c_i f_i = f,$$

and hence trivially $S_n^* f \to f$.

If f is an arbitrary element of $[f_n]$, then given $\varepsilon > 0$, we can find $g = \sum_{i=1}^{m} c_i f_i$ such that $\|f - g\| < \varepsilon/(M+1)$, where $M = \sup_n \|S_n\| < \infty$. It follows that for all $n \geqq m$

$$\begin{aligned}
\|S_n^* f - f\| &\leqq \|S_n^* f - S_n^* g\| + \|S_n^* g - g\| + \|g - f\| \\
&= \|S_n^* f - S_n^* g\| + \|f - g\| \\
&\leqq (\|S_n\| + 1)\|f - g\| < \varepsilon,
\end{aligned}$$

so that once again $S_n^* f \to f$. Thus every element of $[f_n]$ has at least one representation of the form $f = \sum_{n=1}^{\infty} c_n f_n$. Since $f(x_n) = c_n$ for every n, the representation is unique and $\{f_n\}$ is a basis for its closed linear span. ■

When X is a *reflexive* Banach space, $\{f_n\}$ is a fortiori complete.

Theorem 6. *If $\{x_n\}$ is a basis for a reflexive Banach space X, then the associated sequence of coefficient functionals $\{f_n\}$ is a basis for X^*.*

Proof. In view of Theorem 5, we need only establish that $\{f_n\}$ is complete in X^*. Suppose then that $\mu \in X^{**}$ and that $\mu(f_n) = 0$ for $n = 1, 2, 3, \ldots$. It is to be shown that $\mu = 0$. Let P be the canonical imbedding of X into X^{**}, i.e.,

$$(Px)(f) = f(x)$$

for f in X^* and x in X. Since X is reflexive, it follows that $\mu = Px$ for some x; accordingly, $f_n(x) = 0$ for $n = 1, 2, 3, \ldots$. Since $\{x_n\}$ is a basis, we have $x = \sum_{n=1}^{\infty} f_n(x)x_n = 0$. Thus $\mu = 0$. ■

For the remainder of this section X will denote a fixed Hilbert space.

Two sequences $\{x_n\}$ and $\{y_n\}$ in X are said to be **biorthogonal** if

$$(x_m, y_n) = \delta_{mn}$$

for every m and n. The Hahn-Banach theorem shows that for a given sequence $\{x_n\}$ a biorthogonal sequence $\{y_n\}$ will exist if and only if $\{x_n\}$ is *minimal*[†] (this means that each element of the sequence lies outside the closed linear span of the others). If this condition is fulfilled, then the biorthogonal sequence $\{y_n\}$ will be uniquely determined if and only if $\{x_n\}$ is complete.

Suppose now that X is separable and that $\{x_n\}$ is a basis for X. The isomorphism between X and X^* shows that to each coefficient functional f_n there corresponds an element $y_n \in X$ such that $f_n(x) = (x, y_n)$ for all x. Since $f_n(x_m) = \delta_{mn}$, we see that $\{x_n\}$ and $\{y_n\}$ are biorthogonal.

Thus we have shown that every basis $\{x_n\}$ for a Hilbert space possesses a unique biorthogonal sequence $\{y_n\}$. In terms of this biorthogonal pair, each vector x in the space can be uniquely represented in the form

$$x = \sum_{n=1}^{\infty} (x, y_n)x_n.$$

Combining Theorems 5 and 6, we see that $\{y_n\}$ is also a basis for X and that, by duality,

$$x = \sum_{n=1}^{\infty} (x, x_n)y_n.$$

Thus, *in a Hilbert space the sequence biorthogonal to a basis is itself a basis.*

PROBLEMS

1. Let $\{x_n\}$ be a basis for a Banach space X and let $\{f_n\}$ be the associated sequence of coefficient functionals. Prove or disprove: if X^* is separable, then $\{f_n\}$ is complete (and hence a basis) in X^*.
2. A sequence $\{x_n\}$ of elements of a Banach space X is said to have $\{f_n\}$, where $f_n \in X^*$, as a **biorthogonal sequence** if $f_i(x_j) = \delta_{ij}$ for every i and j.
   (a) Show that $\{x_n\}$ has a biorthogonal sequence if and only if it is minimal.
   (b) Show that a biorthogonal sequence for $\{x_n\}$ is uniquely determined if and only if $\{x_n\}$ is complete in X.

[†] The terms *(topologically) independent* and *(topologically) free* are also used.

(c) Show that $\{x_n\}$ is a basis for X if and only if it possesses a biorthogonal sequence $\{f_n\}$ such that $\sum_{n=1}^{\infty} f_n(x)x_n$ converges to x for every x in X.

3. **(Karlin)** Let $\{x_n\}$ be a sequence of elements of a Banach space X and let $\{f_n\}$ be biorthogonal to $\{x_n\}$. Prove that if $\{f_n\}$ is a basis for X^*, then $\{x_n\}$ is a basis for X.

4. Show by an example that in a Hilbert space, a sequence biorthogonal to a complete sequence need not itself be complete.

8 RIESZ BASES

The simplest and perhaps the most obvious way of constructing new bases from old is through an isomorphism of the underlying space. Thus, if $\{x_n\}$ is a fixed but arbitrary basis for a Banach space X and if the bounded invertible operator† T transforms $\{x_n\}$ into $\{y_n\}$, that is, if

$$Tx_n = y_n \quad \text{for} \quad n = 1, 2, 3, \ldots,$$

then $\{y_n\}$ is also a basis for X. In fact, it is easy to see that $\{x_n\}$ and $\{y_n\}$ are equivalent in the following sense.

Definition. *Two bases $\{x_n\}$ and $\{y_n\}$ for a Banach space X are said to be **equivalent** if*

$$\sum_{n=1}^{\infty} c_n x_n \text{ is convergent if and only if } \sum_{n=1}^{\infty} c_n y_n \text{ is convergent.}$$

The property of being linked by an isomorphism of the space completely characterizes equivalent bases.

Theorem 7. *Two bases $\{x_n\}$ and $\{y_n\}$ for a Banach space X are equivalent if and only if there exists a bounded invertible operator $T: X \to X$ such that $Tx_n = y_n$ for every n.*

Proof. The sufficiency is trivial. Suppose then that $\{x_n\}$ and $\{y_n\}$ are equivalent bases for X. If $x \in X$, with $x = \sum_{n=1}^{\infty} c_n x_n$, then the series $\sum_{n=1}^{\infty} c_n y_n$ converges to an element Tx in X. The function T thus defined is clearly linear, one-to-one and onto, and $Tx_n = y_n$ for every n.

To show that T is a bounded invertible operator, we define functions T_n by setting $T_n x = \sum_{i=1}^{n} c_i y_i$. Then

$$Tx = \lim_{n\to\infty} T_n x$$

† "Invertible" means one-to-one *and* onto.

for every x. Since each T_n is bounded (by Theorem 3), it follows that T is bounded (by the Banach–Steinhaus theorem). The open mapping theorem guarantees that T is invertible, and the result follows. ∎

Theorem 8. *In a Hilbert space equivalent bases have equivalent biorthogonal sequences.*

Proof. Let $\{x_n\}$ and $\{y_n\}$ be equivalent bases for a Hilbert space H and let $\{f_n\}$ and $\{g_n\}$ be their respective biorthogonal sequences. Observe to begin with that a sequence biorthogonal to a basis is also a basis, so that $\{f_n\}$ and $\{g_n\}$ are themselves bases for H.

Let T be a bounded invertible operator on H such that $Tx_n = y_n$ $(n = 1, 2, 3, \ldots)$. Then the adjoint operator T^* is also bounded and invertible. Assertion: $T^*g_n = f_n$ $(n = 1, 2, 3, \ldots)$. Indeed, for fixed n and all values of m we have

$$(T^*g_n, x_m) = (g_n, Tx_m) = (g_n, y_m) = \delta_{nm} = (f_n, x_m);$$

since $\{x_m\}$ is complete, the assertion follows. This proves that $\{f_n\}$ and $\{g_n\}$ are equivalent bases for H. ∎

In a separable Hilbert space the most important bases are orthonormal. Second in importance are those bases that are equivalent to some orthonormal basis. They will be called *Riesz* bases, and they constitute the largest and most tractable class of bases known.

Definition. *A basis for a Hilbert space is a* ***Riesz basis*** *if it is equivalent to an orthonormal basis, that is, if it is obtained from an orthonormal basis by means of a bounded invertible operator.*

A Riesz basis $\{f_n\}$ for a Hilbert space is necessarily **bounded,** that is to say,

$$0 < \inf_n \|f_n\| \leqq \sup_n \|f_n\| < \infty.$$

In fact, if $\{f_n\}$ is obtained from the orthonormal basis $\{e_n\}$ by means of the bounded invertible operator T, then for every n

$$\frac{1}{\|T^{-1}\|} \leqq \|f_n\| \leqq \|T\|.$$

It follows readily from this that if $\{f_n\}$ is a Riesz basis, then the sequence of unit vectors $\{f_n/\|f_n\|\}$ is also a Riesz basis. Reason: There exists a bounded invertible operator S for which

$$Se_n = \frac{e_n}{\|f_n\|} \quad (n = 1, 2, 3, \ldots);$$

therefore, the operator TS is bounded and invertible, and

$$(TS)e_n = \frac{f_n}{\|f_n\|} \quad (n = 1, 2, 3, \ldots).$$

The next theorem provides a number of important characteristic properties of Riesz bases.

Theorem 9. *Let H be a separable Hilbert space. Then the following statements are equivalent.*

(1) The sequence $\{f_n\}$ forms a Riesz basis for H.

(2) There is an equivalent[†] *inner product on H, with respect to which the sequence $\{f_n\}$ becomes an orthonormal basis for H.*

(3) The sequence $\{f_n\}$ is complete in H, and there exist positive constants A and B such that for an arbitrary positive integer n and arbitrary scalars $c_1, \ldots, c_n$ one has

$$A \sum_{i=1}^{n} |c_i|^2 \leqq \left\| \sum_{i=1}^{n} c_i f_i \right\|^2 \leqq B \sum_{i=1}^{n} |c_i|^2.$$

(4) The sequence $\{f_n\}$ is complete in H, and its Gram matrix

$$((f_i, f_j))_{i,j=1}^{\infty}$$

generates a bounded invertible operator on l^2.

(5) The sequence $\{f_n\}$ is complete in H and possesses a complete biorthogonal sequence $\{g_n\}$ such that

$$\sum_{n=1}^{\infty} |(f, f_n)|^2 < \infty \quad \textit{and} \quad \sum_{n=1}^{\infty} |(f, g_n)|^2 < \infty$$

for every f in H.

Proof. (1) $\Rightarrow$ (2): Since $\{f_n\}$ is a Riesz basis for H, there exists a bounded invertible operator T that transforms $\{f_n\}$ into some orthonormal basis $\{e_n\}$, i.e.,

$$Tf_n = e_n \quad \text{for } n = 1, 2, 3, \ldots.$$

Define a new inner product $(f, g)_1$ on H by setting

$$(f, g)_1 = (Tf, Tg),$$

[†] Two inner products are said to be **equivalent** if they generate equivalent norms.

and let $\|\ \|_1$ be the norm generated by this inner product. Then

$$\frac{\|f\|}{\|T^{-1}\|} \leqq \|f\|_1 \leqq \|T\| \cdot \|f\|$$

for every f in H, so that the new inner product is equivalent to the original one. Clearly,

$$(f_i, f_j)_1 = (Tf_i, Tf_j) = (e_i, e_j) = \delta_{ij}$$

for every i and j.

(2) $\Rightarrow$ (3): Suppose that $(f, g)_1$ is an equivalent inner product on H relative to which the sequence $\{f_n\}$ forms an orthonormal basis. From the relations

$$m\|f\| \leqq \|f\|_1 \leqq M\|f\| \quad \text{for every } f,$$

where m and M are positive constants not depending on f, it follows that for arbitrary scalars $c_1, \ldots, c_n$ one has

$$\frac{1}{M^2}\sum_{i=1}^{n}|c_i|^2 \leqq \left\|\sum_{i=1}^{n} c_i f_i\right\|^2 \leqq \frac{1}{m^2}\sum_{i=1}^{n}|c_i|^2.$$

Clearly, the sequence $\{f_n\}$ is complete in H.

(3) $\Rightarrow$ (1): Let $\{e_n\}$ be an arbitrary orthonormal basis for H. It follows by assumption that there exist bounded linear operators T and S such that $Te_n = f_n$ and $Sf_n = e_n$ $(n = 1, 2, 3, \ldots)$. Certainly, $ST = I$. Since $\{f_n\}$ is complete, we also have $TS = I$. Hence T is invertible, and $\{f_n\}$ is a Riesz basis for H.

(1) $\Rightarrow$ (4): Let T be a bounded invertible operator on H that carries some orthonormal basis $\{e_n\}$ into the basis $\{f_n\}$. If $A = (a_{ij})$ denotes the matrix of the (invertible) operator T^*T relative to $\{e_n\}$, then

$$a_{ij} = (T^*Te_j, e_i) = (Te_j, Te_i) = (f_j, f_i).$$

Therefore, the Gram matrix of $\{f_n\}$ is the *conjugate* of A.

(4) $\Rightarrow$ (3): Suppose that the Gram matrix of $\{f_n\}$ generates a bounded invertible operator on l^2. If $\{e_n\}$ is an arbitrary orthonormal basis for H, then the transformation $T: H \to H$, defined by

$$T\left(\sum_i c_i e_i\right) = \sum_i e_i\left(\sum_j \overline{(f_i, f_j)}c_j\right)$$

whenever $\{c_i\} \in l^2$, is obviously linear, bounded, and invertible. A straightforward calculation shows that

$$\left\|\sum_i c_i f_i\right\|^2 = \left(T\left(\sum_i c_i e_i\right), \sum_i c_i e_i\right),$$

so that, in addition, T is a positive operator. (Recall that a bounded linear operator T on a Hilbert space is said to be *positive* if $(Tf, f) \geqq 0$ for every vector f in the space.) Since T is positive, it has a (unique) positive square root (see Riesz and Sz.-Nagy [1955, p. 265]); call it P. The equation above may then be put in the form

$$\left\|\sum_i c_i f_i\right\|^2 = \left\|P\left(\sum_i c_i e_i\right)\right\|^2,$$

from which it follows at once that

$$\frac{1}{\|T^{-1}\|}\sum_i |c_i|^2 \leqq \left\|\sum_i c_i f_i\right\|^2 \leqq \|T\|\sum_i |c_i|^2.$$

(1) $\Rightarrow$ (5): Let $\{g_n\}$ be the unique sequence in H biorthogonal to $\{f_n\}$. By Theorem 8, $\{g_n\}$ is also a Riesz basis for H. Since every vector f in H has the two biorthogonal expansions

$$f = \sum_{n=1}^{\infty}(f, g_n)f_n \quad \text{and} \quad f = \sum_{n=1}^{\infty}(f, f_n)g_n,$$

the result follows immediately from the definition of a Riesz basis.

(5) $\Rightarrow$ (1): Consider the linear transformation from H into l^2 defined by

$$f \to \{(f, f_n)\}.$$

The reader will verify without great difficulty that this mapping is closed. By the closed graph theorem it is continuous, and hence there exists a positive constant C for which

$$\sum_{n=1}^{\infty}|(f, f_n)|^2 \leqq C^2\|f\|^2 \quad \text{for all } f.$$

Similarly, there exists a positive constant D for which

$$\sum_{n=1}^{\infty}|(f, g_n)|^2 \leqq D^2\|f\|^2 \quad \text{for all } f.$$

Fix an arbitrary orthonormal basis $\{e_n\}$ for H, and define operators S and T on the linear subspaces spanned by the sequences $\{f_n\}$ and $\{g_n\}$, respectively, by setting

$$S\left(\sum_i c_i f_i\right) = \sum_i c_i e_i \quad \text{and} \quad T\left(\sum_i c_i g_i\right) = \sum_i c_i e_i.$$

By virtue of the two inequalities above, we have

$$\left\| S\left(\sum_i c_i f_i\right)\right\| \leqq D \left\|\sum_i c_i f_i\right\|$$

and

$$\left\| T\left(\sum_i c_i g_i\right)\right\| \leqq C \left\|\sum_i c_i g_i\right\|.$$

Since both sequences $\{f_n\}$ and $\{g_n\}$ are complete, each of the operators S and T can be extended by continuity to a bounded linear operator on the entire space. If $f = \Sigma a_i f_i$ and $g = \Sigma b_j g_j$ are *finite* sums, a simple calculation shows that

$$(Sf, Tg) = (f, g);$$

by continuity, this holds for *every* pair of vectors f and g. We have, accordingly,

$$(f, S^*Tg) = (f, g),$$

so that $S^*T = I$. The existence of a right-inverse for S^* implies that S^* is onto, and hence that S is bounded from below (see Taylor [1958, p. 234]). Since the range of S is dense in H, we conclude that S is invertible. Thus the sequence $\{f_n\}$ forms a Riesz basis for H. ■

The class of Riesz bases is very large. It is extremely difficult to exhibit at least one bounded basis for a Hilbert space that is not equivalent to an orthonormal basis. We mention without proof the following example of such a basis in the space $L^2[-\pi, \pi]$; it was discovered by Babenko [1948].

Example. The sequence $\{|t|^\alpha e^{int}\}_{n=-\infty}^{\infty}$, with $0 < \alpha < \frac{1}{2}$, is a bounded basis for $L^2[-\pi, \pi]$ that is not a Riesz basis. The appearance of simplicity is misleading; the example is exceedingly difficult.

PROBLEMS

1. Prove that a sequence that is biorthogonal to a Riesz basis is also a Riesz basis.
2. Suppose that $\{f_n\}$ is a sequence of vectors in a Hilbert space H such that $\Sigma|(f, f_n)|^2 < \infty$ whenever $f \in H$. Show that the mapping $f \to \{(f, f_n)\}$ has a closed graph.
3. Let $\{f_n\}$ be a Riesz basis for a Hilbert space H and let $\{g_n\}$ be biorthogonal to $\{f_n\}$. Theorem 9 guarantees that there exist positive constants A and B such that

$$A\sum|(f, g_n)|^2 \leqq \|f\|^2 \leqq B\sum|(f, g_n)|^2 \quad \text{for all } f.$$

Show that one has the dual relation

$$\frac{1}{B}\sum |(f, f_n)|^2 \leqq \|f\|^2 \leqq \frac{1}{A}\sum |(f, f_n)|^2 \quad \text{for all } f.$$

4. Let $\{f_n\}$ be a basis for a Hilbert space H and let $\{g_n\}$ be biorthogonal to $\{f_n\}$. We shall call $\{f_n\}$ a **Bessel basis** if

$$\sum_{n=1}^{\infty} c_n f_n \quad \text{is convergent only if} \quad \sum_{n=1}^{\infty} |c_n|^2 < \infty;$$

we shall call $\{f_n\}$ a **Hilbert basis** if

$$\sum_{n=1}^{\infty} c_n f_n \quad \text{is convergent if} \quad \sum_{n=1}^{\infty} |c_n|^2 < \infty.$$

(a) Show that $\{f_n\}$ is a Riesz basis if and only if it is both a Bessel basis and a Hilbert basis.

(b) Show that $\{f_n\}$ is a Bessel (Hilbert) basis if and only if $\{g_n\}$ is a Hilbert (Bessel) basis.

(c) Show that $\{f_n\}$ is a Bessel basis if and only if there exists a constant $A > 0$ such that

$$A\sum_{i=1}^{n} |c_i|^2 \leqq \left\|\sum_{i=1}^{n} c_i f_i\right\|^2$$

for arbitrary scalars $c_1, \ldots, c_n$ $(n = 1, 2, 3, \ldots)$.

(d) Show that $\{f_n\}$ is a Hilbert basis if and only if there exists a constant $B > 0$ such that

$$\left\|\sum_{i=1}^{n} c_i f_i\right\|^2 \leqq B\sum_{i=1}^{n} |c_i|^2$$

for arbitrary scalars $c_1, \ldots, c_n$ $(n = 1, 2, 3, \ldots)$.

5. Let $\{f_n\}$ be a basis for a Hilbert space H and let $\{g_n\}$ be its biorthogonal basis. Prove that $\{f_n\}$ and $\{g_n\}$ are equivalent if and only if $\{f_n\}$ is a Riesz basis.

6. Let $\{f_n\}$ be a basis for a Hilbert space H and let $g_n = \lambda_n f_n$, where

$$0 < \inf_n |\lambda_n| \leqq \sup_n |\lambda_n| < \infty.$$

Prove or disprove: $\{f_n\}$ and $\{g_n\}$ are equivalent.

9 THE STABILITY OF BASES IN BANACH SPACES

Two mathematical objects that are in some sense "close" to each other often share common properties. In this and the remaining section of Chapter I, we will show that bases in Banach spaces form a *stable* class in the sense that sequences sufficiently close to bases are themselves bases.

Part of the problem is to formulate broad and effective notions of "close". Let $\{x_n\}$ be a basis for a Banach space X and let $\{y_n\}$ be a sequence of elements in X. When shall we say that $\{y_n\}$ is "close" to $\{x_n\}$? Although there are many different criteria, they all have one element in common: each implies that the mapping

$$x_n \to y_n \quad \text{for} \quad n = 1, 2, 3, \ldots$$

can be extended to an isomorphism T from X onto X that is in some sense "close" to the identity operator I. In this way, questions about the stability of bases can be reduced to questions about "small" perturbations of the identity operator. As we shall see, the operator approach provides a powerful tool in the solution of stability problems.

The fundamental stability criterion, and historically the first, is due to Paley and Wiener [1934]. It is based on the elementary fact that a bounded linear operator T on a Banach space is invertible whenever

$$\|I - T\| < 1.$$

(This is one of those striking instances in which linear operators behave like ordinary numbers: if $|1 - t| < 1$, then surely t^{-1} exists.)

Theorem 10 (Paley–Wiener). *Let $\{x_n\}$ be a basis for a Banach space X, and suppose that $\{y_n\}$ is a sequence of elements of X such that*

$$\left\| \sum_{i=1}^{n} c_i(x_i - y_i) \right\| \leqq \lambda \left\| \sum_{i=1}^{n} c_i x_i \right\|$$

for some constant λ, $0 \leqq \lambda < 1$, and all choices of the scalars $c_1, \ldots, c_n$ $(n = 1, 2, 3, \ldots)$. Then $\{y_n\}$ is a basis for X equivalent to $\{x_n\}$.

Proof. It follows by assumption that the series $\sum_{n=1}^{\infty} c_n(x_n - y_n)$ is convergent whenever the series $\sum_{n=1}^{\infty} c_n x_n$ is convergent. Define a mapping $T{:}\, X \to X$ by setting

$$T\left(\sum_{n=1}^{\infty} c_n x_n\right) = \sum_{n=1}^{\infty} c_n(x_n - y_n).$$

Evidently T is linear and bounded and $\|T\| \leqq \lambda < 1$. Thus the operator $I - T$ is invertible. Since $(I - T)x_n = y_n$ for every n, the result follows. ■

Corollary. *Let $\{x_n\}$ be a basis for a Banach space X and let $\{f_n\}$ be the associated sequence of coefficient functionals. If $\{y_n\}$ is a sequence of vectors* *in X for which*

$$\sum_{n=1}^{\infty} \|x_n - y_n\| \cdot \|f_n\| < 1,$$

then $\{y_n\}$ *is a basis for X equivalent to* $\{x_n\}$.

Proof. Put $\lambda = \sum_{n=1}^{\infty} \|x_n - y_n\| \cdot \|f_n\|$. If $x = \sum c_i x_i$ is an arbitrary *finite* sum, then

$$\left\|\sum c_i(x_i - y_i)\right\| = \left\|\sum f_i(x)(x_i - y_i)\right\| \leqq \sum \|f_i(x)(x_i - y_i)\|$$
$$\leqq \lambda\|x\| = \lambda \left\|\sum c_i x_i\right\|.$$

Since $0 \leqq \lambda < 1$, the result follows from Theorem 10. ■

The following theorem is now immediate.

Theorem 11 (Krein–Milman–Rutman). *If* $\{x_n\}$ *is a basis for a Banach space X, then there exist numbers* $\varepsilon_n > 0$ *with the following property: if* $\{y_n\}$ *is a sequence of vectors in X for which*

$$\|x_n - y_n\| < \varepsilon_n, \quad n = 1, 2, 3, \ldots,$$

then $\{y_n\}$ *is a basis for X equivalent to* $\{x_n\}$.

Proof. Let $\{f_n\}$ be the sequence of coefficient functionals associated with the basis $\{x_n\}$. By the corollary to Theorem 10, it suffices to choose ε_n small enough so that $\sum_{n=1}^{\infty} \varepsilon_n \|f_n\| < 1$. ■

In the following two corollaries the strength of the Krein–Milman–Rutman theorem is forcefully revealed.

Corollary 1. *If X is a Banach space with a basis, then every dense subset of X contains a basis.*

Corollary 2. *The space* $C[a, b]$ *has a basis consisting entirely of polynomials.*

It is important for applications that the corollary to Theorem 10 be strengthened. For this purpose, we shall make use of a well-known result concerning "compact perturbations" of the identity operator. If T is a compact operator defined on a Banach space X and if $\text{Ker}(I - T) = \{0\}$, then $I - T$ is invertible.

This is the *Fredholm alternative* (see Halmos [1967, Problem 140]); it lends support to the notion that a compact operator is in some sense "small".

Definition. *A sequence $\{x_n\}$ of elements of a Banach space X is said to be* ***ω-independent*** *if the equality*

$$\sum_{n=1}^{\infty} c_n x_n = 0$$

is possible only for $c_n = 0$ $(n = 1, 2, 3, \ldots)$.

Theorem 12. *Let $\{x_n\}$ be a basis for a Banach space X and let $\{f_n\}$ be the associated sequence of coefficient functionals. If $\{y_n\}$ is complete in X and if*

$$\sum_{n=1}^{\infty} \|x_n - y_n\| \cdot \|f_n\| < \infty,$$

then $\{y_n\}$ is a basis for X equivalent to $\{x_n\}$.

Proof. Let us first show that the sequence $\{y_n\}$ is ω-independent. If it were not, then we could find scalars $c_1, c_2, c_3, \ldots$ (not all zero) such that $\sum_{n=1}^{\infty} c_n y_n = 0$. Choose N so large that $\sum_{n=N+1}^{\infty} \|x_n - y_n\| \cdot \|f_n\| < 1$. It follows by the corollary to Theorem 10 that the sequence of vectors

$$\{x_1, \ldots, x_N, y_{N+1}, y_{N+2}, \ldots\}$$

forms a basis for X. Accordingly, at least one of the scalars $c_1, \ldots, c_N$ does not vanish; suppose $c_k \neq 0$. The equation $\sum_{n=1}^{\infty} c_n y_n = 0$ can then be solved for y_k, showing that $X = [y_n] = [y_n]_{n \neq k}$. Since $1 \leqq k \leqq N$, it follows that the codimension[†] of $[y_n]_{n>N}$ is at most $N - 1$. But this is absurd: the quotient space $X/[y_n]_{n>N}$ is isomorphic to $[x_1, \ldots, x_N]$, and so has dimension N. This proves that $\{y_n\}$ is ω-independent.

Define an operator T on X by setting

$$Tx = \sum_{n=1}^{\infty} f_n(x)(x_n - y_n).$$

It is clear that T is a bounded linear operator. We claim that T is compact. Indeed, if we put $T_n x = \sum_{i=1}^{n} f_i(x)(x_i - y_i)$, then by hypothesis $\|T - T_n\| \to 0$, so that T is the limit of operators of finite rank, and hence compact.

We complete the proof by showing that $\operatorname{Ker}(I - T) = \{0\}$. If $(I - T)x = 0$, then $\sum_{n=1}^{\infty} f_n(x) y_n = 0$ and since $\{y_n\}$ is ω-independent, it follows that $x = 0$.

[†] If Y is a closed subspace of a Banach space X, then the *codimension* of Y is defined to be the dimension of X/Y.

Thus the kernel of $I - T$ is trivial. The proof is over: the Fredholm alternative shows that $I - T$ is invertible and clearly $(I - T)x_n = y_n$ for every n. ■

PROBLEMS

1. Show that if $\lambda = 1$, then Theorem 10 is no longer valid.
2. Suppose that $\{x_n\}$ is a complete sequence of vectors belonging to a Banach space X and that $\left\|\sum c_i(x_i - y_i)\right\| \leqq \lambda \left\|\sum c_i x_i\right\|$ for some constant λ, $0 \leqq \lambda < 1$, and arbitrary scalars $c_1, \ldots, c_n$ $(n = 1, 2, 3, \ldots)$. Show that $\{y_n\}$ is also complete.
3. **(Retherford–Holub)** Let $\{x_n\}$ be a basis for a Hilbert space H with $\|x_n\| = 1$ for every n, and let $\{y_n\}$ be biorthogonal to $\{x_n\}$. Show that if $\sum_{n=1}^{\infty} \sqrt{\|y_n\| - 1} < \infty$, then $\{x_n\}$ is a Riesz basis. (*Hint*: Use Theorem 12 to show that $\{y_n\}$ is equivalent to $\{x_n\}$, and then use Problem 5 of Section 8.)

10 THE STABILITY OF ORTHONORMAL BASES IN HILBERT SPACE

Throughout this section H will denote a separable Hilbert space and $\{e_n\}$ a fixed but arbitrary orthonormal basis for H. While it is of course true that every result of the preceding section applies equally well to orthonormal bases, the added structure of a Hilbert space provides additional and stronger stability criteria.

We begin by reformulating Theorem 10.

Theorem 13. *Let $\{e_n\}$ be an orthonormal basis for a Hilbert space H and let $\{f_n\}$ be "close" to $\{e_n\}$ in the sense that*

$$\left\|\sum c_i(e_i - f_i)\right\| \leqq \lambda \sqrt{\sum |c_i|^2}$$

for some constant λ, $0 \leqq \lambda < 1$, and arbitrary scalars $c_1, \ldots, c_n$ $(n = 1, 2, 3, \ldots)$. Then $\{f_n\}$ is a Riesz basis for H.

APPLICATION: THE STABILITY OF THE TRIGONOMETRIC SYSTEM

It follows from what has already been done that the trigonometric system $\{e^{int}\}_{-\infty}^{\infty}$ is *stable* in $L^2[-\pi, \pi]$ under "sufficiently small" perturbations of the integers. This means that if $\{\lambda_n\}$ is a sequence of real or complex numbers for which $\{\lambda_n - n\}$ is in some sense "small", then the system $\{e^{i\lambda_n t}\}_{-\infty}^{\infty}$ will form a basis for $L^2[-\pi, \pi]$, in fact, a Riesz basis. Accordingly, every function f in

$L^2[-\pi, \pi]$ will have a unique **nonharmonic Fourier series** expansion

$$f(t) = \sum_{-\infty}^{\infty} c_n e^{i\lambda_n t} \quad \text{(in the mean)},$$

with $\sum |c_n|^2 < \infty$.

The possibility of such *nonharmonic* expansions was discovered by Paley and Wiener [1934], and it was for this purpose that they formulated the criterion of Theorem 13. In the present setting that criterion takes the form

$$\left\| \sum c_n (e^{int} - e^{i\lambda_n t}) \right\| \leqq \lambda < 1$$

whenever $\sum |c_n|^2 \leqq 1$.

When shall the sequence $\{\lambda_n - n\}$ be considered "small"? Based on what has already been established, one might well suppose that the condition

$$\lambda_n - n \to 0 \quad \text{as} \quad n \to \pm\infty$$

is, at the very least, necessary. Surprisingly, it is not.

Theorem 14 (Kadec's $\frac{1}{4}$-Theorem). *If $\{\lambda_n\}$ is a sequence of real numbers for which*

$$|\lambda_n - n| \leqq L < \tfrac{1}{4}, \quad n = 0, \pm 1, \pm 2, \ldots,$$

then $\{e^{i\lambda_n t}\}$ satisfies the Paley–Wiener criterion and so forms a Riesz basis for $L^2[-\pi, \pi]$.

Proof. It is to be shown that

$$\left\| \sum c_n (e^{int} - e^{i\lambda_n t}) \right\| \leqq \lambda < 1$$

whenever $\sum |c_n|^2 \leqq 1$. Write

$$e^{int} - e^{i\lambda_n t} = e^{int}(1 - e^{i\delta_n t}),$$

where $\delta_n = \lambda_n - n$ $(n = 0, \pm 1, \pm 2, \ldots)$. The trick is to expand the function $1 - e^{i\delta t} (-\pi \leqq t \leqq \pi)$ in a Fourier series relative to the complete orthonormal system $\{1, \cos nt, \sin(n - \frac{1}{2})t\}_{n=1}^{\infty}$ (see Problem 9, Section 3) and then exploit the fact that $|\lambda_n - n|$ is not too large.

Simple calculations show that when δ is real,

$$1 - e^{i\delta t} = \left(1 - \frac{\sin \pi\delta}{\pi\delta}\right) + \sum_{k=1}^{\infty} \frac{(-1)^k 2\delta \sin \pi\delta}{\pi(k^2 - \delta^2)} \cos kt$$

$$+ i \sum_{k=1}^{\infty} \frac{(-1)^k 2\delta \cos \pi\delta}{\pi((k - \frac{1}{2})^2 - \delta^2)} \sin\left(k - \frac{1}{2}\right)t.$$

Let $\{c_n\}$ be an arbitrary *finite* sequence of scalars such that $\sum |c_n|^2 \leqq 1$. By interchanging the order of summation and then using the triangle inequality we see that

$$\left\|\sum c_n(e^{int} - e^{i\lambda_n t})\right\| \leqq A + B + C,$$

where

$$A = \left\|\sum_n \left(1 - \frac{\sin \pi\delta_n}{\pi\delta_n}\right) c_n e^{int}\right\|,$$

$$B = \sum_{k=1}^{\infty} \left\|\cos kt \sum_n \frac{(-1)^k 2\delta_n \sin \pi\delta_n}{\pi(k^2 - \delta_n^2)} c_n e^{int}\right\|,$$

and

$$C = \sum_{k=1}^{\infty} \left\|\sin\left(k - \frac{1}{2}\right)t \sum_n \frac{(-1)^k 2\delta_n \cos \pi\delta_n}{\pi((k-\frac{1}{2})^2 - \delta_n^2)} c_n e^{int}\right\|.$$

Obvious estimates show that

$$A \leqq \left(1 - \frac{\sin \pi L}{\pi L}\right), \quad B \leqq \sum_{k=1}^{\infty} \frac{2L \sin \pi L}{\pi(k^2 - L^2)}, \quad C \leqq \sum_{k=1}^{\infty} \frac{2L \cos \pi L}{\pi((k-\frac{1}{2})^2 - L^2)}.$$

But the series $\sum_{k=1}^{\infty} 2L/\pi(k^2 - L^2)$ and $\sum_{k=1}^{\infty} 2L/\pi((k-\frac{1}{2})^2 - L^2)$ are the partial fraction expansions of the functions $1/\pi L - \cot \pi L$ and $\tan \pi L$, respectively (see Markushevich [1965, pp. 62, 64]), so that

$$\left\|\sum c_n(e^{int} - e^{i\lambda_n t})\right\| \leqq \lambda = 1 - \cos \pi L + \sin \pi L.$$

The proof is over: $L < \frac{1}{4}$ implies $\lambda < 1$. ■

Remarks.

1. The result is sharp in the sense that the constant $\frac{1}{4}$ cannot be improved. In fact, if $L = \frac{1}{4}$, then the conclusion of the theorem no longer holds. A counterexample is provided by the sequence $\{\lambda_n\}$, with

$$\lambda_n = \begin{cases} n - \frac{1}{4}, & n > 0, \\ 0, & n = 0, \\ n + \frac{1}{4}, & n < 0. \end{cases}$$

Writing it down is easy; proving that it works is another matter (and hard). We defer the proof until Section 3.3.

2. The proof of Theorem 14 applies only when every λ_n is real. Even earlier, however, it had been shown by Duffin and Eachus [1942] that the

Paley–Wiener criterion is satisfied whenever the λ_n are complex and

$$|\lambda_n - n| \leqq L < \frac{\log 2}{\pi}, \qquad n = 0, \pm 1, \pm 2, \ldots.$$

The method of proof is the same, only now the function $1 - e^{i\delta t}$ is expanded in an everywhere-convergent Taylor series (see Problem 2). Whether the constant $(\log 2)/\pi$ can be replaced by $\frac{1}{4}$ (for complex λ_n) remains an unsolved problem.

Any further analysis of nonharmonic Fourier series in $L^2[-\pi, \pi]$ requires the use of deep structural properties of entire functions. These will be discussed in Chapter 2. At present we can go no further.

Let us return then to the abstract setting. It follows at once from Theorem 13 that if

$$\sum_{n=1}^{\infty} \|e_n - f_n\|^2 < 1,$$

then $\{f_n\}$ is a basis for H equivalent to $\{e_n\}$, that is, a Riesz basis. If $\{f_n\}$ is known to be ω-independent, then more can be said. We first give a definition.

Definition. *Two sequences of vectors $\{f_n\}$ and $\{g_n\}$ in a normed vector space are said to be* ***quadratically close*** *if*

$$\sum_{n=1}^{\infty} \|f_n - g_n\|^2 < \infty.$$

Theorem 15 (Bari). *Let H be a separable Hilbert space and $\{e_n\}$ an orthonormal basis for H. If $\{f_n\}$ is an ω-independent sequence that is quadratically close to $\{e_n\}$, then $\{f_n\}$ is a Riesz basis for H.*

Proof. Define an operator $T: H \to H$ by setting

$$Tf = \sum_{n=1}^{\infty} (f, e_n)(e_n - f_n).$$

It is clear that T is linear and that

$$\|T\|^2 \leqq \sum_{n=1}^{\infty} \|e_n - f_n\|^2.$$

Furthermore, since $Te_n = e_n - f_n$, it follows that

$$\sum_{n=1}^{\infty} \|Te_n\|^2 = \sum_{n=1}^{\infty} \|e_n - f_n\|^2 < \infty.$$

This shows that T is a *Hilbert–Schmidt operator* and hence compact (see Halmos [1967, Problem 135]).

We complete the proof by showing that $\operatorname{Ker}(I - T) = \{0\}$. If $(I - T)f = 0$, then from the equations

$$0 = (I - T)f = \sum_n (f, e_n)e_n - \sum_n (f, e_n)(e_n - f_n) = \sum_n (f, e_n)f_n$$

and the fact that $\{f_n\}$ is ω-independent, it follows that $f = 0$. Hence $\operatorname{Ker}(I - T) = \{0\}$, and the Fredholm alternative shows that $I - T$ is invertible. Clearly $(I - T)e_n = f_n$ for every n. ■

A basis that is quadratically close to an orthonormal basis is called a *Bari basis*. Bari bases form an important subclass of the class of all Riesz bases. An excellent discussion of some of their special properties, together with applications to the theory of non-selfadjoint operators, can be found in Gohberg and Krein [1969] (also see Problems 5–7 at the end of this section).

Since every orthonormal sequence of vectors is ω-independent, the following result is an immediate consequence of Theorem 15.

Theorem 16 (Birkhoff–Rota). *Let $\{e_n\}$ be an orthonormal basis for a Hilbert space H. If $\{f_n\}$ is an orthonormal sequence that is quadratically close to $\{e_n\}$, then $\{f_n\}$ is complete, and hence an orthonormal basis.*

Remark. It is possible to give an elementary proof of Theorem 16. In this context, a proof is "elementary" if it depends only on the intrinsic geometry of the Hilbert space and not on operator theory. The reader will appreciate the power of the operator approach by tackling the "elementary" proof sketched in the exercises (see Problem 4).

APPLICATION: EIGENFUNCTION EXPANSIONS

An important class of boundary-value problems of mathematical physics, including the classical vibration problems of continuum mechanics (vibrating strings, membranes, and elastic bars, for example) can be reduced to the (regular) *Sturm–Liouville system*

$$u'' + (\lambda - q(t))u = 0 \tag{1}$$

($q(t)$ is assumed to be of class C^1 on $[a, b]$), together with the *separated* endpoint conditions

$$\alpha u(a) + u'(a) = 0, \quad \beta u(b) + u'(b) = 0. \tag{2}$$

The following facts about such a system are well known (see Birkhoff and Rota [1962]).

1. There is a discrete set of real eigenvalues $\lambda = \lambda_n$ $(n = 0, 1, 2, \ldots)$, and $\lambda_n \to \infty$ as $n \to \infty$;

2. Eigenfunctions associated with different eigenvalues are orthogonal in $L^2[a, b]$; and

3. The corresponding *normalized* eigenfunctions are given by the asymptotic formulas

$$u_n(t) = \sqrt{2}\cos\left(\frac{n\pi(t-a)}{b-a}\right) + \frac{\delta_n(t)}{n+1} \qquad (n = 0, 1, 2, \ldots),$$

where $\delta_n(t)$ is a bounded function of n and t.

Accordingly, the solutions to the system above behave in much the same way as the solutions to the system

$$u'' + \lambda u = 0,$$

with the endpoint conditions $u'(a) = u'(b) = 0$.

It is now a simple matter to show that the eigenfunctions associated with any Sturm–Liouville system of the form (1), (2) are always complete in $L^2[a, b]$.

Theorem 17. *If* u_n $(n = 0, 1, 2, \ldots)$ *denotes the* n*th normalized eigenfunction of the system (1), (2), then the* u_n*'s form an orthonormal basis for* $L^2[a, b]$.

Proof. In $L^2[a, b]$, the functions e_n defined by $e_0(t) = 1$ and

$$e_n(t) = \sqrt{2}\cos\left(\frac{n\pi(t-a)}{b-a}\right) \quad \text{for} \quad a \leqq t \leqq b \quad (n = 1, 2, 3, \ldots)$$

form an orthonormal basis. The asymptotic formula for u_n shows that

$$\sum_{n=0}^{\infty} \|e_n - u_n\|^2 < \infty,$$

and the result follows at once from Theorem 16. ■

Consequently, every function f in $L^2[a, b]$ can be expressed in terms of the eigenfunctions; the corresponding series

$$f(t) = \sum_{n=0}^{\infty} c_n u_n(t)$$

is called a *Sturm–Liouville series.*

We conclude our discussion of stability by further clarifying the relation between Riesz bases and the Paley–Wiener criterion. As always, H is a separable Hilbert space and $\{e_n\}$ an orthonormal basis for H.

The Paley–Wiener criterion is nothing more than the assertion that the mapping

$$T: e_n \to f_n \quad \text{for} \quad n = 1, 2, 3, \ldots$$

can be extended to an isomorphism on all of H for which $\|I - T\| < 1$. This is a stringent requirement to place on a linear operator, and one might well conclude that the Paley–Wiener theory is of very limited scope. In a sense, just the opposite is true: every Riesz basis for H is obtained in *essentially* this way.

Theorem 18. *Let H be a separable Hilbert space and $\{f_n\}$ a Riesz basis for H. Then there exists an orthonormal basis $\{e_n\}$, an isomorphism T, and a positive number ρ such that*

$$Te_n = \rho f_n \quad \textit{for} \quad n = 1, 2, 3, \ldots$$

and

$$\|I - T\| < 1.$$

The proof will require the following two well-known facts (see Riesz and Sz.-Nagy [1955, p. 230] for the first and Halmos [1967, Problem 105] for the second).

Lemma 1. *If T is a bounded self-adjoint operator on a Hilbert space, then*

$$\|T\| = \sup\left\{\frac{|(Tf, f)|}{\|f\|^2} : f \neq 0\right\}.$$

Lemma 2 (Polar Decomposition). *Every bounded invertible operator T on a Hilbert space can be factored in the form $T = UP$, where U is a unitary operator and P is a positive operator.*

Proof of Theorem 18. Since $\{f_n\}$ is a Riesz basis for H, there exist positive constants A and B such that

$$A\sum |c_n|^2 \leqq \left\|\sum c_n f_n\right\|^2 \leqq B\sum |c_n|^2$$

whenever $\{c_n\} \in l^2$. Choose

$$\rho = \frac{2}{\sqrt{A} + \sqrt{B}}$$

and set $g_n = \rho f_n$ $(n = 1, 2, 3, \ldots)$. The inequalities above may then be put in the form

$$(1 - \lambda)\sqrt{\sum |c_n|^2} \leqq \left\|\sum c_n g_n\right\| \leqq (1 + \lambda)\sqrt{\sum |c_n|^2},$$

where

$$\lambda = \frac{\sqrt{B} - \sqrt{A}}{\sqrt{B} + \sqrt{A}} < 1.$$

We complete the proof by showing that there exists an orthonormal basis $\{e_n\}$ such that

$$\left\| \sum c_n(e_n - g_n) \right\| \leqq \lambda \sqrt{\sum |c_n|^2}.$$

Select an *arbitrary* orthonormal basis $\{\phi_n\}$ for H and define a mapping $S: H \to H$ by putting

$$S\left(\sum c_n \phi_n\right) = \sum c_n g_n$$

whenever $\{c_n\} \in l^2$. Then S is a bounded linear operator and

$$(1 - \lambda)\|f\| \leqq \|Sf\| \leqq (1 + \lambda)\|f\|$$

for every f in H. This shows, in particular, that S is bounded from below. Since $\{g_n\}$ is complete, S has a dense range and so must be invertible.

Let $S = UP$ be the polar decomposition of S. Then

$$\|Pf\| \leqq (1 + \lambda)\|f\| \quad \text{whenever} \quad f \in H,$$

and by Lemma 1,

$$(Pf, f) \leqq (1 + \lambda)\|f\|^2.$$

Rewriting this as

$$(Pf - f, f) \leqq \lambda\|f\|^2,$$

and again using Lemma 1, we find

$$\|f - Pf\| \leqq \lambda\|f\| \quad \text{for all } f.$$

Define $e_n = U\phi_n$ $(n = 1, 2, 3, \ldots)$. Since U is unitary, the sequence $\{e_n\}$ forms an orthonormal basis for H. If $\{c_n\}$ is an arbitrary square-summable sequence of scalars and if we put $f = \sum c_n \phi_n$, then

$$\begin{aligned} \left\| \sum c_n(e_n - g_n) \right\| &= \|Uf - Sf\| = \|U(f - Pf)\| \\ &= \|f - Pf\| \leqq \lambda\|f\| = \lambda \sqrt{\sum |c_n|^2}. \end{aligned}$$

The proof is complete since the mapping $T: \sum_{n=1}^{\infty} c_n e_n \to \sum_{n=1}^{\infty} c_n g_n$ has all the desired properties. ∎

PROBLEMS

1. Let $\{e_n\}$ be an orthonormal basis for H and let $\{f_n\}$ be a sequence of vectors in H such that

$$\left\|\sum c_n(e_n - f_n)\right\| < \sqrt{\sum |c_n|^2}$$

whenever $0 < \sum |c_n|^2 < \infty$. Show that $\{f_n\}$ is complete. Must it be a basis?

2. (Duffin–Eachus) Show that the system of exponentials $\{e^{i\lambda_n t}\}_{n=-\infty}^{\infty}$ forms a Riesz basis for $L^2[-\pi, \pi]$ whenever

$$|\lambda_n - n| \leqq L < \frac{\log 2}{\pi}, \qquad n = 0, \pm 1, \pm 2, \ldots.$$

(*Hint*: Show that the Paley–Wiener criterion is satisfied by expanding the function $1 - e^{i\delta t}$ in an everywhere-convergent Taylor series and then mimicking the proof of Theorem 14.)

3. (Schäfke) Let $\{e_n\}$ be an orthonormal basis for H and let $\{f_n\}$ be "close" to $\{e_n\}$ in the sense that

$$\sum_{n=1}^{\infty} |(f, e_n - f_n)|^2 \leqq \lambda < 1 \quad \text{whenever} \quad \|f\| \leqq 1.$$

Show that $\{f_n\}$ is a Riesz basis for H. (*Hint*: Define

$$Tf = \sum_{n=1}^{\infty} (f, e_n - f_n)e_n$$

and then show that $T^*f = \sum_{n=1}^{\infty} (f, e_n)(e_n - f_n)$.)

4. *Alternative proof of Theorem 16.*

(a) Choose N large enough so that $\sum_{n=N+1}^{\infty} \|e_n - f_n\|^2 < 1$. Then the system

$$\{e_1, \ldots, e_N, f_{N+1}, f_{N+2}, \ldots\}$$

is complete in H.

(b) Define

$$g_n = e_n - \sum_{k=N+1}^{\infty} (e_k, f_k)f_k \quad \text{for} \quad n = 1, \ldots, N.$$

Then

$$\{g_1, \ldots, g_N, f_{N+1}, f_{N+2}, \ldots\}$$

is also complete in H.

(c) Let S be the orthogonal complement of the set $\{f_{N+1}, f_{N+2}, f_{N+3}, \ldots\}$. Then $S = [g_1, \ldots, g_N]$.

(d) Finally, $S = [f_1, \ldots, f_N]$. Therefore, if $f \perp f_n$ $(n = 1, 2, 3, \ldots)$, then $f = 0$.

5. If $\{f_n\}$ is a Bari basis for H, then so is $\{f_n / \|f_n\|\}$.

6. In order that a sequence of vectors $\{f_n\}$ be a Bari basis for H, it is necessary and sufficient that there exist an orthonormal basis $\{e_n\}$ and an invertible operator T on H such that
(1) $Te_n = f_n$ for $n = 1, 2, 3, \ldots$, and
(2) $I - T$ is a Hilbert–Schmidt operator.

7. Let $\{f_n\}$ be a basis for H that is quadratically close to the orthonormal basis $\{e_n\}$ and let $\{g_n\}$ be the basis biorthogonal to $\{f_n\}$. Show that $\{g_n\}$ is also quadratically close to $\{e_n\}$ and hence that the bases $\{f_n\}$ and $\{g_n\}$ are quadratically close.

2

ENTIRE FUNCTIONS OF EXPONENTIAL TYPE

There is an intimate connection between analytic functions and the completeness of sets of complex exponentials $\{e^{i\lambda_n t}\}$. If, for example, the set $\{e^{i\lambda_n t}\}$ fails to be complete in $C[a, b]$, then by virtue of the Riesz representation theorem, there is a function $\omega(t)$ of bounded variation on the interval $[a, b]$ that is not essentially a constant and for which

$$\int_a^b e^{i\lambda_n t}\, d\omega(t) = 0 \quad (n = 1, 2, 3, \ldots).$$

If we let $f(z)$ be the Fourier–Stieltjes transform of ω, i.e., if

$$f(z) = \int_a^b e^{izt}\, d\omega(t),$$

then $f(z)$ is an entire function, not identically zero, and $f(z)$ vanishes at every λ_n. In this way the study of the completeness properties of the system $\{e^{i\lambda_n t}\}$ reduces to the study of the zeros of certain entire functions.

The representation above for an entire function $f(z)$ places a severe restriction on its growth. On the real axis, $f(z)$ evidently is bounded, while off the axis, it can grow no faster than an exponential:

$$|f(z)| \leqq Ae^{B|z|}.$$

An entire function (or any function analytic in a sector) satisfying such an inequality for suitable constants A and B is said to be of *exponential type*. Familiar examples include e^z, $\sin z$, $\cos\sqrt{z}$, as well as all polynomials.

Any limitation on the growth of an entire function carries with it limitations on the distribution of its zeros. In fact, the fundamental question in the theory of entire functions is precisely that of the connection between growth and distribution of zeros. The more zeros a polynomial has, for example, the higher its degree and hence the more rapid its growth.

The relation between the growth of an entire function and the distribution of its zeros was investigated by Borel, Hadamard, Lindelöf, and others in the late nineteenth and early twentieth centuries. The basis for this investigation is the classical theorem of Weierstrass on the expansion of entire functions into infinite products. Accordingly, we begin with the Weierstrass factorization theorem.

PART ONE. THE CLASSICAL FACTORIZATION THEOREMS

1 WEIERSTRASS'S FACTORIZATION THEOREM

Every polynomial can be written as a product of linear factors. Entire functions behave in much the same way: if $f(z)$ is entire and has only finitely many zeros, $z_1, \dots, z_n$ (we shall always assume that a zero of order k is repeated k times), then we can "factor out" the zeros and obtain

$$f(z) = (z - z_1)\cdots(z - z_n)g(z),$$

where $g(z)$ is entire and *never* zero. Any attempt to extend this process to the case of infinitely many zeros encounters serious convergence problems; a more subtle approach is therefore needed.

Let $z_1, z_2, z_3, \dots$ be an infinite sequence of complex numbers, none of which is zero, with $\lim z_n = \infty$. We are going to construct an entire function with precisely these zeros. The most natural choice for such a function is the infinite product

$$\prod_{n=1}^{\infty}\left(1 - \frac{z}{z_n}\right).$$

Observe, however, that unless the series $\sum_{n=1}^{\infty} 1/|z_n|$ is convergent, the product cannot converge absolutely (except for $z = 0$) and therefore *may not* represent an entire function. What is needed, in general, are "convergence-producing" factors.

We are going to show that there exist polynomials $p_n(z)$ such that the product

$$\prod_{n=1}^{\infty}\left(1-\frac{z}{z_n}\right)e^{p_n(z)} \tag{1}$$

converges to an entire function uniformly on each bounded region of the plane, and that $p_n(z)$ can be chosen to be

$$p_n(z)=\sum_{k=1}^{n}\frac{1}{k}\left(\frac{z}{z_n}\right)^k \qquad (n=1,2,3,\ldots). \tag{2}$$

For this purpose we introduce the Weierstrass **primary factors**

$$E(u,0)=1-u$$

and

$$E(u,p)=(1-u)\exp\left(u+\frac{1}{2}u^2+\cdots+\frac{1}{p}u^p\right) \quad \text{for} \quad p=1,2,3,\ldots.$$

Note that when $|u|<1$,

$$\log E(u,p)=-\sum_{k=p+1}^{\infty}\frac{u^k}{k}$$

(the *principal value* of the logarithm is to be chosen), so that when $|u|\leqq\varepsilon<1$,

$$|\log E(u,p)|\leqq\sum_{k=p+1}^{\infty}|u|^k\leqq\frac{|u|^{p+1}}{1-\varepsilon}. \tag{3}$$

It follows easily from this estimate that the series

$$\sum_{n=1}^{\infty}\log E\left(\frac{z}{z_n},n\right)$$

converges uniformly on each bounded region that contains none of the points z_n. Accordingly, if $p_n(z)$ is given by (2), then the infinite product (1) converges uniformly on *every* bounded region of the plane and so represents an entire function with zeros at each z_n and at these points only.

It is now a simple matter to show that every entire function can be factored in this way.

Theorem 1 (Weierstrass). *Every entire function $f(z)$ that is not identically zero can be represented in the form*

$$f(z)=z^m e^{g(z)}\prod_{n=1}^{\infty}\left(1-\frac{z}{z_n}\right)e^{p_n(z)},$$

where the product is taken over all the zeros of $f(z)$ other than $z = 0$, m is a nonnegative integer, $g(z)$ is an entire function, and the $p_n(z)$ are given by (2).

Proof. Let $z_1, z_2, z_3, \ldots$ be the zeros of $f(z)$ other than $z = 0$, and suppose that $f(z)$ has a zero of order m at the origin. In the case of infinitely many zeros, it is clear that $\{z_n\}$ can have no finite limit point, and hence $\lim z_n = \infty$.

If we define

$$\phi(z) = z^m \prod_{n=1}^{\infty} \left(1 - \frac{z}{z_n}\right) e^{p_n(z)},$$

with $p_n(z)$ given by (2), then $\phi(z)$ is entire and has the same zeros as $f(z)$. Therefore, $f(z)/\phi(z)$ is entire and never zero, so that

$$\frac{f(z)}{\phi(z)} = e^{g(z)}$$

for some entire function $g(z)$. The result follows. ■

The usefulness of the Weierstrass factorization theorem is limited by the fact that the polynomials $p_n(z)$ are of such large degree. There is, however, one case in which the expansion of $f(z)$ can be greatly simplified. Let us suppose that for some nonnegative integer p,

$$\sum_{n=1}^{\infty} \frac{1}{|z_n|^{p+1}} < \infty.$$

The estimate (3) can once again be applied, and we conclude in the same way as before that the product

$$P(z) = \prod_{n=1}^{\infty} E\left(\frac{z}{z_n}, p\right)$$

converges uniformly (and absolutely) on each bounded region of the plane. Therefore, in this case, $f(z)$ can be written as

$$f(z) = z^m e^{g(z)} P(z).$$

If p is the smallest nonnegative integer for which the series $\sum_{n=1}^{\infty} 1/|z_n|^{p+1}$ is convergent, then $P(z)$ is called the **canonical product** associated with the sequence $\{z_n\}$, and the number p is called the **genus** of the canonical product. We shall show subsequently that a canonical factorization is always possible provided that $f(z)$ is not of too rapid growth.

Example. As an application, we obtain the canonical expansion of $\sin \pi z$. Here there is a simple zero at each integer: $z_n = n$ $(n = 0, \pm 1, \pm 2, \ldots)$. Since

$\sum_{n=1}^{\infty} 1/n$ diverges while $\sum_{n=1}^{\infty} 1/n^2$ converges, the genus of the corresponding canonical product is $p = 1$, and we have

$$\sin \pi z = z e^{g(z)} \prod_{n \neq 0} \left(1 - \frac{z}{n}\right) e^{z/n}.$$

To determine $g(z)$, take the logarithmic derivative of both sides. Then

$$\pi \cot \pi z = \frac{1}{z} + g'(z) + \sum_{n \neq 0} \left(\frac{1}{z-n} + \frac{1}{n}\right).$$

But

$$\pi \cot \pi z = \frac{1}{z} + \sum_{n \neq 0} \left(\frac{1}{z-n} + \frac{1}{n}\right),$$

so that $g'(z) = 0$ and hence $g(z)$ is a constant. Since $\lim_{z \to 0} (\sin \pi z)/z = \pi$, it follows that $e^{g(z)} = \pi$, and hence

$$\sin \pi z = \pi z \prod_{n \neq 0} \left(1 - \frac{z}{n}\right) e^{z/n}.$$

Now, the product converges absolutely for all values of z, so that rearrangement is permissible, and we obtain the simpler formula

$$\sin \pi z = \pi z \prod_{n=1}^{\infty} \left(1 - \frac{z^2}{n^2}\right).$$

PROBLEMS

1. Prove that for $|z| < 1$,

$$(1+z)(1+z^2)(1+z^4)(1+z^8)\cdots = \frac{1}{1-z}.$$

2. Show that when $z_n = i/n$ $(n = 1, 2, 3, \ldots)$, the product $\prod(1+z_n)$ diverges while the product $\prod |1+z_n|$ converges.

3. Exhibit a sequence $\{z_n\}$ of real numbers such that $\sum z_n$ converges and $\prod(1+z_n)$ diverges.

4. Suppose that $\sum |z_n|^2 < \infty$. Show that the product $\prod(1+z_n)$ is convergent if and only if the series $\sum z_n$ is convergent.

5. Show that the value of an absolutely convergent product remains the same if its factors are reordered.

6. If the product $\prod(1 - z/z_n)$ is absolutely convergent for all values of z, then it represents an entire function. What if it converges *conditionally* for at least one value of z?

7. Find an infinite product expansion for each of the following entire functions: (a) $\cos z$, (b) $\cosh z$, (c) $\sinh z$, (d) $e^z - 1$, (e) $e^{az} - e^{bz}$.

8. Derive **Wallis's product:** $\dfrac{\pi}{2} = \dfrac{2}{1}\dfrac{2}{3}\dfrac{4}{3}\dfrac{4}{5}\dfrac{6}{5}\dfrac{6}{7}\cdots$.

9. Show that

$$\sin(z+a) = (\sin a)e^{z\cot a}\prod_{n=-\infty}^{\infty}\left(1+\frac{z}{a+n\pi}\right)e^{-z/(a+n\pi)}$$

provided that a is not a multiple of π.

10. Show that every function that is meromorphic in the entire plane is the quotient of two entire functions.

11. Let $z_1, z_2, z_3, \ldots$ be a sequence of distinct complex numbers with $\lim z_n = \infty$ and let $w_1, w_2, w_3, \ldots$ be arbitrary complex numbers. Then there exists an entire function $f(z)$ such that $f(z_n) = w_n$ $(n = 1, 2, 3, \ldots)$. (*Hint*: Show that if $g(z)$ has a simple zero at each z_n, then there exist constants $c_1, c_2, c_3, \ldots$ such that the series

$$\sum_{n=1}^{\infty}\frac{w_n g(z)}{g'(z_n)(z-z_n)}e^{c_n(z-z_n)}$$

converges uniformly in every bounded region of the plane.)

12. Let $z_1, z_2, z_3, \ldots$ be a sequence of distinct complex numbers with $\lim z_n = \infty$ and let $w_1, w_2, w_3, \ldots$ be arbitrary complex numbers. Show that there exists an entire function $f(z)$ such that

$$\int_{z_n}^{z_{n+1}} f(z)\,dz = w_n \quad (n = 1, 2, 3, \ldots).$$

2 JENSEN'S FORMULA

In this section we shall establish a relation between the modulus of an analytic function on a circle and the moduli of its zeros inside the circle. It is known as *Jensen's formula*, and it is one of the most important theorems in analysis.

Theorem 2 (Jensen's Formula). *Let $f(z)$ be analytic in $|z| < R$ and suppose that $f(0) \neq 0$. If $z_1, \ldots, z_n$ are the zeros of $f(z)$ in $|z| \leqq r\,(0 < r < R)$, then*

$$\frac{1}{2\pi}\int_0^{2\pi}\log|f(re^{i\theta})|\,d\theta = \log|f(0)| + \sum_{k=1}^{n}\log\left(\frac{r}{|z_k|}\right).$$

Proof. The result is immediate when $f(z)$ is never zero in $|z| \leqq r$, for then $\log|f(z)|$ is harmonic and the mean-value property gives

$$\frac{1}{2\pi}\int_0^{2\pi} \log|f(re^{i\theta})|\,d\theta = \log|f(0)|. \tag{1}$$

The essence of the proof is the observation that this formula remains valid even when $f(z)$ has zeros lying on the circle $|z| = r$ (but not in its interior).

Suppose then that $z_1, \ldots, z_n$ all lie on $|z| = r$, and write $z_k = re^{i\theta_k}$ ($k = 1, \ldots, n$). There is but one thing to do in the presence of unwanted zeros: we eliminate them. Define

$$g(z) = f(z)\prod_{k=1}^{n} \frac{z_k}{z_k - z}.$$

Then $g(z)$ is analytic and free of zeros in $|z| \leqq r$. Replacing f by g in (1), we obtain

$$\frac{1}{2\pi}\int_0^{2\pi} \log|f(re^{i\theta})|\,d\theta = \log|f(0)| + \sum_{k=1}^{n}\frac{1}{2\pi}\int_0^{2\pi} \log|1 - e^{i(\theta-\theta_k)}|\,d\theta. \tag{2}$$

Using the calculus of residues, it is not hard to show that

$$\int_0^{\pi} \log\sin\theta\,d\theta = -\pi\log 2$$

(see Ahlfors [1979, p. 160]), and this implies

$$\int_0^{2\pi} \log|1 - e^{i\theta}|\,d\theta = 0.$$

It follows that each of the integrals on the right-hand side of (2) must vanish, and hence (1) remains valid.

The general case is now readily established. Given arbitrary zeros $z_1, \ldots, z_n$ inside or on the circle $|z| = r$, we form the function

$$F(z) = f(z)\prod_{k=1}^{n} \frac{r^2 - \bar{z}_k z}{r(z - z_k)}.$$

Then $F(z)$ is analytic and free of zeros in $|z| < r$. Since $|F(z)| = |f(z)|$ whenever $|z| = r$, it follows from what has already been proved that

$$\log|F(0)| = \frac{1}{2\pi}\int_0^{2\pi} \log|f(re^{i\theta})|\,d\theta.$$

But

$$|F(0)| = |f(0)|\prod_{k=1}^{n}\left(\frac{r}{|z_k|}\right),$$

and the result follows. ∎

It is frequently useful to state Jensen's formula in a slightly altered form. If $f(z)$ is analytic in $|z| < R$, then we denote by $n(r)$, $0 \leqq r < R$, the number of zeros $z_1, z_2, z_3, \ldots$ of $f(z)$ for which $|z_n| \leqq r$. Provided that $f(0) \neq 0$, it follows easily that

$$\sum_{|z_n| \leqq r} \log\left(\frac{r}{|z_n|}\right) = \int_0^r \frac{n(t)}{t}\,dt,$$

and Jensen's formula becomes

$$\frac{1}{2\pi}\int_0^{2\pi} \log|f(re^{i\theta})|\,d\theta = \log|f(0)| + \int_0^r \frac{n(t)}{t}\,dt.$$

Jensen's formula provides a powerful tool for studying the relation between the growth of an entire function and the density of its zeros: the slower the growth, the more sparsely distributed the zeros. As an application of this principle, we shall apply Jensen's formula to the class of entire functions of exponential type.

Definition. *An entire function $f(z)$ is said to be of* ***exponential type*** *if the inequality*

$$|f(z)| \leqq Ae^{B|z|}$$

holds for some positive constants A and B and all values of z.

Theorem 3. *If $f(z)$ is an entire function of exponential type, then $n(r)/r$ remains bounded as $r \to \infty$.*

Proof. We may suppose without loss of generality that $f(0) = 1$. This is obvious if $f(0) \neq 0$; if, on the other hand, $f(z)$ has a zero of order m at the origin, then we need only consider $f(z)/z^m$.

Set

$$N(r) = \int_0^r \frac{n(t)}{t}\,dt.$$

With this notation, Jensen's formula becomes

$$N(r) = \frac{1}{2\pi}\int_0^{2\pi} \log|f(re^{i\theta})|\,d\theta.$$

By assumption, $|f(z)| \leqq Ae^{B|z|}$ for some positive constants A and B and all values of z, so that

$$\log|f(re^{i\theta})| \leqq \log A + Br$$

and hence also

$$N(r) \leqq \log A + Br.$$

Since $n(r)$ is a nondecreasing function of r, it follows that

$$n(r)\log 2 = n(r)\int_r^{2r}\frac{1}{t}\,dt \leqq \int_r^{2r}\frac{n(t)}{t}\,dt \leqq N(2r).$$

Thus, for all values of r,

$$n(r)\log 2 \leqq \log A + 2Br,$$

and the result follows. ■

PROBLEMS

1. Using the calculus of residues show that

$$\int_0^{\pi}\log\sin\theta\,d\theta = -\pi\log 2.$$

2. Prove that if $f(z)$ is entire and $n(r)$ is the number of zeros of $f(z)$ in $|z| \leqq r$, then, even if $f(0) = 0$,

$$\int_1^r \frac{n(t)}{t}\,dt < \frac{1}{2\pi}\int_0^{2\pi}\log|f(re^{i\theta})|\,d\theta + A$$

for some constant A that depends only on $f(z)$ and not on $r (r > 1)$.

3. Calculate $n(r)$ for each of the entire functions $e^z - 1$, $\sin z$, and $(\sin\sqrt{z})/\sqrt{z}$.

4. Obtain an asymptotic estimate for $n(r)$ when

$$f(z) = \prod_{n=1}^{\infty}\left(1 - \frac{z}{n!}\right).$$

3 FUNCTIONS OF FINITE ORDER

To characterize the growth of an entire function $f(z)$, we introduce the "maximum modulus function"

$$M(r) = \max\{|f(z)| : |z| = r\}.$$

Unless $f(z)$ is a constant, $M(r)$ is a strictly increasing function of r, and $\lim_{r\to\infty} M(r) = \infty$. The first assertion follows from the maximum modulus principle, while the second is a consequence of Liouville's theorem.

An entire function that grows no faster than a polynomial must in fact be a polynomial. More precisely, we have the following theorem.

Theorem 4. *If $f(z)$ is an entire function for which*

$$M(r) \leqq r^n$$

for some integer n and all sufficiently large values of r, then $f(z)$ is a polynomial of degree at most n.

Proof. Expand $f(z)$ in a Taylor series,

$$f(z) = \sum_{k=0}^{\infty} a_k z^k,$$

and set

$$p(z) = a_0 + a_1 z + \cdots + a_n z^n.$$

If

$$g(z) = \frac{f(z) - p(z)}{z^{n+1}},$$

then $g(z)$ is entire, and by hypothesis, $g(z) \to 0$ uniformly as $|z| \to \infty$. It follows from the maximum modulus principle that $g(z)$ must vanish identically, so that $f(z)$ is a polynomial of degree no larger than n. ■

Remark. It is evident that the theorem remains valid (and the proof remains the same) under the weaker assumption that $M(r) \leqq r^n$ on a *sequence* of circles $|z| = r_n$ with $r_n \to \infty$.

In light of Theorem 4, to measure the growth of a transcendental entire function, it is necessary to compare $M(r)$ with functions that grow faster than every power of r.

Definition. *An entire function $f(z)$ is said to be of* ***finite order*** *if there exists a positive number k such that*

$$M(r) \leqq e^{r^k}$$

as soon as r is "sufficiently large", i.e., $r > r(k)$. The greatest lower bound of all positive numbers k for which this is true is called the ***order*** *of the function and is denoted by ρ.*

Thus, ρ is the smallest nonnegative number such that

$$M(r) \leqq e^{r^{\rho+\varepsilon}}$$

for every positive value of ε, as soon as r is sufficiently large. It follows easily from the definition that the order of a nonconstant entire function is given by the formula

$$\rho = \limsup_{r\to\infty} \frac{\log\log M(r)}{\log r}.$$

The order of a constant function is of course 0.

Simple examples of functions of finite order include e^z, $\sin z$, and $\cos z$, all of which are of order 1, and $\cos\sqrt{z}$, which is of order $\frac{1}{2}$. An entire function of exponential type is of finite order at most 1; every polynomial is of order 0; the function e^{e^z} is of infinite order.

Theorem 3 is easily modified for functions of finite order (the details are left to the reader).

Theorem 5. *If $f(z)$ is an entire function of finite order ρ, then*

$$n(r) = O(r^{\rho+\varepsilon})$$

for every positive number ε.

As a corollary, we have the following important result.

Theorem 6. *If $f(z)$ is an entire function of finite order ρ and if $z_1, z_2, z_3, \ldots$ are its zeros, other than $z = 0$, then the series*

$$\sum_{n=1}^{\infty} \frac{1}{|z_n|^{\alpha}}$$

is convergent whenever $\alpha > \rho$.

Proof. We may suppose without loss of generality that the zeros of $f(z)$ have been numbered so that

$$0 < |z_1| \leqq |z_2| \leqq \cdots.$$

Given $\alpha > \rho$, choose β so that $\rho < \beta < \alpha$. Since $f(z)$ is of order ρ, it follows from Theorem 5 that

$$n(r) \leqq Ar^{\beta}$$

for some constant A and all values of r. Take $r = |z_n|$; then $n(r) \geqq n$, and hence

$$n \leqq A|z_n|^{\beta} \quad \text{for} \quad n = 1, 2, 3, \ldots.$$

The result follows at once since $\sum_{n=1}^{\infty} 1/n^{\alpha/\beta} < \infty$. ■

It is an important consequence of Theorem 6 that every entire function of finite order has a canonical factorization.

Corollary. *If $f(z)$ is an entire function of finite order ρ, with zeros $z_1, z_2, z_3, \ldots$, other than $z = 0$, then*

$$f(z) = z^m e^{g(z)} \prod_{n=1}^{\infty} E\left(\frac{z}{z_n}, p\right),$$

where the product is a canonical product of genus p and $p \leqq \rho$.

Proof. The remarks following Theorem 1 show that the product $\prod_{n=1}^{\infty} E(z/z_n, p)$ converges uniformly on each bounded region of the plane whenever

$$\sum_{n=1}^{\infty} \frac{1}{|z_n|^{p+1}} < \infty,$$

and this is certainly true as long as $p + 1 > \rho$. ∎

Much more is true. We shall prove subsequently (Theorem 9) that $g(z)$ is a *polynomial* of degree no larger than ρ.

Definition. *Let $z_1, z_2, z_3, \ldots$ be a sequence of complex numbers, none of which is zero. The greatest lower bound of positive numbers α for which the series*

$$\sum_{n=1}^{\infty} \frac{1}{|z_n|^{\alpha}}$$

is convergent is called the ***exponent of convergence*** *of the sequence $\{z_n\}$ and is denoted by λ. If the series is divergent for every $\alpha > 0$, then we set $\lambda = \infty$. For a "finite" sequence, $\lambda = 0$ by definition.*

Example. The exponents of convergence of the sequences

$$\{e^n\}, \quad \{n^{1/\lambda}\}, \quad \text{and} \quad \{\log n\}$$

are 0, λ, and ∞, respectively.

If λ is the exponent of convergence of the sequence $\{z_n\}$, then the series $\sum 1/|z_n|^{\alpha}$ converges if $\alpha > \lambda$ and diverges if $\alpha < \lambda$; if $\alpha = \lambda$, then no conclusion can be drawn. For example, the sequences $\{n^{1/\lambda}\}$ and $\{(n \log^2 n)^{1/\lambda}\}$ have the same exponent of convergence λ, but $\sum 1/|z_n|^{\lambda}$ diverges for the first sequence and converges for the second.

Suppose now that $f(z)$ is entire of finite order ρ. Let λ be the exponent of convergence of its zeros and let p be the genus of the corresponding canonical product. Our results thus far may be summarized by the following inequalities: by Theorem 6,

$$\lambda \leqq \rho \quad \text{(Hadamard)};$$

by the definitions alone,

$$p \leqq \lambda \leqq p+1.$$

If λ is *not* an integer, then $p = [\lambda]$ (here $[x]$ denotes the function "bracket x", the largest integer not exceeding x); if λ is an integer, then there is an ambiguity: $p = \lambda$ when the series $\sum 1/|z_n|^\lambda$ diverges, while $p = \lambda - 1$ when it converges.

PROBLEMS

1. Prove that, for a given function, $M(r)$ is a continuous function of r.

2. Determine $M(r)$ for each of the entire functions e^z, $\sin z$, $\cos z$, and $(\sin\sqrt{z})/\sqrt{z}$.

3. Let $f(z)$ be an entire function with $f(0) = 1$. Prove that

$$n(r) \leqq \log M(er),$$

where $n(r)$ is the number of zeros of $f(z)$ in the closed disk $|z| \leqq r$.

4. Let $f(z)$ be an entire function such that

$$\operatorname{Re} f(z) < M$$

for some constant M and all values of z. Show that $f(z)$ is a constant.

5. Determine the order of each of the entire functions e^{2z}, $(\sin\sqrt{z})/\sqrt{z}$, and $\int_0^1 e^{zt^2}\,dt$.

6. Show that the order of a product of two entire functions does not exceed the larger of the orders of the factors.

7. Prove that $f(z)$ and $f'(z)$ are of the same order.

8. Determine the exponent of convergence of the zeros of $e^{e^z} - 1$.

9. Prove that if $f(z)$ has at least one zero, but is not identically zero, then

$$\lambda = \limsup_{r\to\infty} \frac{\log n(r)}{\log r}.$$

10. If $f(z) = \sum_{n=0}^{\infty} a_n z^n$ is an entire function of order ρ, then

$$\rho = \limsup_{n\to\infty} \frac{n \log n}{\log(1/|a_n|)} \tag{1}$$

(the quotient is taken to be zero if $a_n = 0$). (*Hint*: Suppose first that $\rho < \infty$ and let the right-hand side of (1) be denoted by μ.

(a) $\mu \leqq \rho$.
Using Cauchy's estimate for the Taylor coefficients of $f(z)$, show that for each fixed number k greater than ρ, one has

$$|a_n| \leqq r^{-n} e^{r^k} \quad (n = 0, 1, 2, \ldots)$$

as soon as r is sufficiently large, i.e., $r > r(k)$. Show that the quantity on the right achieves its minimum value when $r = (n/k)^{1/k}$, so that

$$|a_n| \leqq \left(\frac{ek}{n}\right)^{n/k} \quad (n \text{ large}).$$

Conclude that $k \geqq \mu$ and hence that $\rho \geqq \mu$.

(b) $\rho \leqq \mu$.
If $k > \mu$, then

$$0 \leqq \frac{n \log n}{\log(1/|a_n|)} \leqq k$$

for all sufficiently large values of n, say $n > N$, so that

$$|a_n| \leqq \left(\frac{1}{n}\right)^{n/k} \quad \text{for} \quad n > N.$$

Then

$$M(r) \leqq \sum_{n=0}^{\infty} |a_n| r^n \leqq \sum_{n=0}^{N} |a_n| r^n + \sum_{n=N+1}^{\infty} \left(\frac{1}{n}\right)^{n/k} r^n.$$

Write

$$\sum_{n=N+1}^{\infty} \left(\frac{1}{n}\right)^{n/k} r^n = S_1 + S_2,$$

where S_1 contains those terms with $n < (2r)^k$ and S_2 the remaining terms. Show that for all values of r, $S_2 \leqq 1$, while for all positive values of ε,

$$S_1 < r^{(2r)^k} \sum \left(\frac{1}{n}\right)^{n/k} = O(e^{r^{k+\varepsilon}}).$$

Conclude that $\rho \leqq k + \varepsilon$, and hence that $\rho \leqq \mu$.)

11. Show that if the right-hand side of (1) in Problem 10 is finite, then $f(z)$ is an entire function of order ρ.

12. Determine the order of each of the following entire functions:

(a) $\sum_{n=0}^{\infty} \dfrac{z^n}{(n!)^\alpha} \quad (\alpha > 0)$,

(b) $\sum_{n=1}^{\infty} \left(\dfrac{\alpha}{n}\right)^{n/\alpha} z^n \quad (\alpha > 0)$,

(c) $\sum_{n=0}^{\infty} e^{-n^2} z^n$,

(d) $J_k(z) = \sum_{n=0}^{\infty} \dfrac{(-1)^n z^{k+2n}}{2^{k+2n} n!(n+k)!}$ (kth **Bessel function**),

(e) $\sum_{n=0}^{\infty} \dfrac{z^n}{\Gamma(1+\alpha n)} \quad \alpha > 0$ (**Mittag–Leffler function**).

13. Show that if $\varepsilon = 0$, then the conclusion of Theorem 5 no longer holds.

4 ESTIMATES FOR CANONICAL PRODUCTS

Hadamard's inequality $\lambda \leqq \rho$ is valid for all entire functions. For canonical products, λ and ρ are equal.

Theorem 6 (Borel). *The order of a canonical product is equal to the exponent of convergence of its zeros.*

Proof. Let

$$P(z) = \prod_{n=1}^{\infty} E\left(\frac{z}{z_n}, p\right)$$

be a canonical product of genus p and order ρ formed with the zeros $z_1, z_2, z_3, \ldots$, and let λ be the exponent of convergence of these zeros. Since $\lambda \leqq \rho$, it is sufficient to show that $\rho \leqq \lambda$. In order to estimate the growth of $P(z)$, we shall estimate the size of each of its factors.

Fix z and write

$$\log|P(z)| = \left(\sum_{r_n \leqq 2r} + \sum_{r_n > 2r}\right) \log\left|E\left(\frac{z}{z_n}, p\right)\right| = \sum\nolimits_1 + \sum\nolimits_2,$$

where $r = |z|$ and $r_n = |z_n|$. For the second sum $\sum_2$ we use the inequality

$$\log|E(u, p)| \leqq 2|u|^{p+1},$$

valid whenever $|u| \leqq \frac{1}{2}$ (by virtue of (3) of Section 1). We then have

$$\sum\nolimits_2 \leqq 2 \sum_{r_n > 2r} \left(\frac{r}{r_n}\right)^{p+1} = 2r^{p+1} \sum_{r_n > 2r} r_n^{-p-1}.$$

If $\lambda = p + 1$, then

$$\sum\nolimits_2 = O(r^{p+1}) = O(r^{\lambda}).$$

If $\lambda < p + 1$, then $\lambda + \varepsilon < p + 1$ whenever ε is sufficiently small, and hence

$$\begin{aligned} \sum\nolimits_2 &\leqq 2r^{p+1} \sum_{r_n > 2r} r_n^{-p-1} = 2r^{p+1} \sum_{r_n > 2r} r_n^{\lambda+\varepsilon-p-1} r_n^{-\lambda-\varepsilon} \\ &< 2r^{p+1}(2r)^{\lambda+\varepsilon-p-1} \sum_{r_n > 2r} r_n^{-\lambda-\varepsilon} = O(r^{\lambda+\varepsilon}). \end{aligned} \tag{1}$$

For the sum $\sum_1$, suppose first that $p > 0$. Clearly,

$$\log|E(u, p)| \leqq \log|1 - u| + |u| + \frac{|u|^2}{2} + \cdots + \frac{|u|^p}{p}.$$

If $|u| \geqq \frac{1}{2}$, then

$$|u|^k \leqq 2^{p-k}|u|^p \quad (k = 1, \ldots, p),$$

and hence

$$\log|E(u, p)| \leqq \log|1-u| + 2^p|u|^p \leqq 2^{p+1}|u|^p,$$

whenever $p > 0$ and $|u| \geqq \frac{1}{2}$. Therefore, for each $\varepsilon > 0$,

$$\begin{aligned}\sum\nolimits_1 &= O\left(\sum_{r_n \leqq 2r}\left(\frac{r}{r_n}\right)^p\right) = O\left(r^p \sum_{r_n \leqq 2r} r_n^{\lambda+\varepsilon-p} r_n^{-\lambda-\varepsilon}\right) \\ &= O\left(r^p (2r)^{\lambda+\varepsilon-p} \sum_{r_n \leqq 2r} r_n^{-\lambda-\varepsilon}\right) = O(r^{\lambda+\varepsilon}). \qquad (2)\end{aligned}$$

If $p = 0$, then

$$\log|E(u, 0)| = O(|u|^\varepsilon)$$

for every positive number ε, and (2) remains valid when p is replaced by ε.

The proof is over: combining (1) and (2), we see that

$$\log|P(z)| = O(r^{\lambda+\varepsilon})$$

for every $\varepsilon > 0$, so that $\rho \leqq \lambda$. ■

The simplest canonical product is of genus zero:

$$P(z) = \prod_{n=1}^{\infty}\left(1 - \frac{z}{z_n}\right) \quad \text{with} \quad \sum_{n=1}^{\infty} \frac{1}{|z_n|} < \infty.$$

By Borel's theorem, $P(z)$ is of order at most 1. However, a lot more can be said: $P(z)$ is actually an entire function of exponential type. We first state the following definition.

Definition. *Let $f(z)$ be an entire function of exponential type. The* ***exponential type***† *of $f(z)$ is defined to be the number*

$$k = \limsup_{r \to \infty} \frac{\log M(r)}{r}.$$

The zero function has exponential type 0, *by convention.*

† The notion of "exponential type" is not to be confused with that of "type". An entire function of *positive* order ρ is said to be of **type** τ (relative to that order) if

$$\tau = \limsup_{r \to \infty} \frac{\log M(r)}{r^\rho}.$$

Thus, for example, the entire function $(\sin\sqrt{z})/\sqrt{z}$ is of order $\rho = \frac{1}{2}$ and type $\tau = 1$, whereas it is clearly of *exponential type* 0.

According to the definition, k is the smallest nonnegative number such that

$$M(r) \leqq e^{(k+\varepsilon)r}$$

for any given $\varepsilon > 0$ as soon as r is sufficiently large.

The functions e^z, $\sin z$, $\cos z$, and $(\sin z)/z$ are all of exponential type 1. Any entire function of order $\rho < 1$ is of exponential type 0.

Theorem 7. *A canonical product of genus zero is an entire function of exponential type zero.*

Proof. Let $P(z)$ be the canonical product

$$P(z) = \prod_{n=1}^{\infty}\left(1 - \frac{z}{z_n}\right) \quad \text{with} \quad \sum_{n=1}^{\infty}\frac{1}{|z_n|} < \infty.$$

We may suppose that the z_n are numbered in nondecreasing order of magnitude. In this case the sequence $\{1/|z_n|\}$ is nonincreasing, and the convergence of the series $\sum 1/|z_n|$ implies that

$$\frac{n}{|z_n|} \to 0 \quad \text{as} \quad n \to \infty.$$

It follows readily from this that if $n(r)$ denotes the number of points $z_1, z_2, z_3, \ldots$ for which $|z_n| \leqq r$, then

$$\frac{n(r)}{r} \to 0 \quad \text{as} \quad r \to \infty \tag{3}$$

(see Problem 1).

Writing $\sum 1/|z_n|$ as a Stieltjes integral and then integrating by parts, we obtain

$$\sum_{n=1}^{\infty}\frac{1}{|z_n|} = \int_0^{\infty}\frac{dn(t)}{t} = \int_0^{\infty}\frac{n(t)}{t^2}\,dt.$$

Hence the integral

$$\int_0^{\infty}\frac{n(t)}{t^2}\,dt \tag{4}$$

is convergent.

We can now estimate the growth of $P(z)$ as follows: if $r = |z|$, then

$$\begin{aligned}\log|P(z)| &\leqq \sum_{n=1}^{\infty}\log\left(1 + \frac{r}{|z_n|}\right)\\ &= \int_0^{\infty}\log\left(1 + \frac{r}{t}\right)\,dn(t)\end{aligned}$$

$$= \int_0^\infty \frac{r}{t(t+r)} n(t)\,dt \quad \text{(integrate by parts)}$$

$$< \int_0^r \frac{n(t)}{t}\,dt + r\int_r^\infty \frac{n(t)}{t^2}\,dt.$$

From this estimate, together with (3) and the convergence of the integral (4), we readily obtain the asymptotic inequality

$$\log|P(z)| < \varepsilon r$$

for each fixed $\varepsilon > 0$, as soon as r is sufficiently large. This shows that $P(z)$ is of exponential type zero. ■

The next theorem shows that the maximum modulus of $1/P(z)$ is very often of the same order of magnitude as the maximum modulus of $P(z)$.

Theorem 8. *If $P(z)$ is a canonical product of order ρ, then for each $\varepsilon > 0$,*

$$|P(z)| > e^{-r^{\rho+\varepsilon}} \tag{5}$$

on circles $|z| = r$ of arbitrarily large radius.

Proof. Let $P(z)$ be a canonical product of genus p and order ρ, formed with the zeros $z_1, z_2, z_3, \ldots$. Fix h, $h > \rho$, and let D_n denote the disk

$$|z - z_n| \leqq |z_n|^{-h}, \qquad n = 1, 2, 3, \ldots.$$

It is to be shown that (5) holds for each $\varepsilon > 0$ whenever z lies outside every disk D_n and $r = |z|$ is sufficiently large, i.e., $r > r(\varepsilon, h)$. Since the sum of the radii of these excluded disks is *finite*, the domain $\mathbf{C} - \cup D_n$ is non-empty and certainly contains circles $|z| = r$ of arbitrarily large radius.

Put $r_n = |z_n|$. The same method used to prove Theorem 6 shows that for every $\varepsilon > 0$,

$$\log|P(z)| \geqq \sum_{r_n \leqq 2r} \log\left|1 - \frac{z}{z_n}\right| - O\left(\sum_{r_n \leqq 2r}\left(\frac{r}{r_n}\right)^p\right) - O\left(\sum_{r_n > 2r}\left(\frac{r}{r_n}\right)^{p+1}\right)$$

$$= \sum_{r_n \leqq 2r} \log\left|1 - \frac{z}{z_n}\right| - O(r^{\rho+\varepsilon}) \tag{6}$$

(remember that $\rho = \lambda$). If z lies outside every excluded disk and if $r_n \leqq 2r$, then

$$\left|1 - \frac{z}{z_n}\right| \geqq r_n^{-h-1} \geqq (2r)^{-h-1}.$$

Therefore, for each $\varepsilon > 0$ and all sufficiently large values of r,

$$\begin{aligned}\sum_{r_n \leqq 2r} \log\left|1 - \frac{z}{z_n}\right| &\geqq -(h+1)\cdot\log(2r)\cdot n(2r)\\ &\geqq -(h+1)\cdot\log(2r)\cdot r^{\rho+\varepsilon} \quad \text{(by Theorem 5)}\\ &> -r^{\rho+2\varepsilon}. \qquad (7)\end{aligned}$$

Combining (6) and (7), we have the desired result. ■

PROBLEMS

1. Prove: if $\{\lambda_n\}$ is a nondecreasing sequence of positive numbers and if $n(r)$ denotes the number of λ_n not exceeding r, then the statements $\lim_{n\to\infty} n/\lambda_n = L$ and $\lim_{r\to\infty} n(r)/r = L$ are equivalent.
2. Determine the order of each of the canonical products

$$\prod_{n=1}^{\infty}\left(1 + \frac{z}{e^n}\right) \quad \text{and} \quad \prod_{n=1}^{\infty}\left(1 + \frac{z}{n^\alpha}\right), \quad \text{with} \quad \alpha > 1.$$

3. Prove that $f(z) = \sum_{n=0}^{\infty} a_n z^n$ is an entire function of exponential type k if and only if

$$k = \limsup_{n\to\infty} \sqrt[n]{n!|a_n|} < \infty.$$

4. Prove: $f(z)$ and $f'(z)$ are of the same exponential type.
5. Give an example of an entire function of order 1 that is not of exponential type.
6. Let $r_1, r_2, r_3, \ldots$ be a sequence of positive real numbers. Show that for each positive number α, the series $\sum_{n=1}^{\infty} 1/r_n^\alpha$ and the integral $\int_0^\infty n(t)/t^{\alpha+1}\,dt$ converge or diverge together.

5 HADAMARD'S FACTORIZATION THEOREM

Equipped with the results of the preceding section, we can now establish the fundamental factorization theorem for entire functions of finite order. It is due to Hadamard who used the result in his celebrated proof of the Prime Number Theorem. It is one of the classical theorems in function theory.

Theorem 9 (Hadamard). *If $f(z)$ is an entire function of finite order ρ and if*

$$f(z) = z^m e^{g(z)} P(z)$$

is its canonical factorization, then $g(z)$ is a polynomial of degree no larger than ρ.

Proof. Let λ denote the exponent of convergence of the zeros of the canonical product $P(z)$. Then $P(z)$ is of order λ and $\lambda \leqq \rho$. Let ε be an arbitrary positive number. By assumption,

$$\log |f(z)| < r^{\rho+\varepsilon}$$

as soon as $|z| = r$ is sufficiently large, while, by Theorem 8,

$$\log |P(z)| > -r^{\lambda+\varepsilon} > -r^{\rho+\varepsilon}$$

on circles $|z| = r$ of arbitrarily large radius. Combining these two inequalities, we see that

$$\operatorname{Re} g(z) = \log \left| \frac{f(z)}{z^m P(z)} \right| < 2r^{\rho+\varepsilon}$$

on a sequence of circles $|z| = r_n$ with $r_n \to \infty$. Since $g(z)$ is entire, the Borel–Carathéodory inequality (see Titchmarsh [1939, p. 174]) shows that

$$|g(z)| = O(r^{\rho+\varepsilon})$$

on the same sequence of circles $|z| = r_n$. Conclusion: $g(z)$ is a polynomial of degree at most $\rho + \varepsilon$ (see the *Remark* following Theorem 4). Since ε was arbitrary, the result follows. ■

Remarks.

1. $\rho = \max(\lambda, \deg g)$.

Indeed, we now know that both λ and the degree of $g(z)$ do not exceed ρ, so that $\max(\lambda, \deg g) \leqq \rho$. But the order of a product of two entire functions cannot exceed the larger of the orders of the factors (Problem 6, Section 3). Since $z^m P(z)$ is of order λ and $e^{g(z)}$ is of order equal to the degree of $g(z)$, we have $\rho \leqq \max(\lambda, \deg g)$, and the result follows.

2. If ρ is not an integer, then $\lambda = \rho$ (for in this case the degree of $g(z)$ is strictly smaller than ρ), and hence the form of the canonical product is uniquely determined: the genus p is equal to $[\rho]$. If, on the other hand, ρ is an integer, then there is an ambiguity: p may be equal to ρ or $\rho - 1$. For example, the canonical products

$$\prod_{n=1}^{\infty} \left(1 - \frac{z}{n}\right) e^{z/n} \quad \text{and} \quad \prod_{n=2}^{\infty} \left(1 - \frac{z}{n(\log n)^2}\right)$$

are both of order 1, but of different genus.

The following impressive corollary is a direct consequence of Hadamard's theorem.

Corollary. *An entire function of nonintegral order assumes every finite value infinitely many times.*

Proof. Since $f(z)$ and $f(z) - c$ have the same order for each constant c, it is enough to show that an entire function of nonintegral order has infinitely many zeros. But this is immediate from *Remark* 2 above since, in this case, the exponent of convergence must be positive. ■

Example 1. Consider the entire function

$$f(z) = \frac{\sin \pi\sqrt{z}}{\pi\sqrt{z}}.$$

It is of order $\rho = \frac{1}{2}$ and has a simple zero at $z = n^2$ $(n = 1, 2, 3, \ldots)$. Therefore, by Hadamard's theorem,

$$f(z) = c \prod_{n=1}^{\infty} \left(1 - \frac{z}{n^2}\right).$$

Since $f(0) = 1$, it follows that $c = 1$. Observe that when z is replaced by z^2, we obtain the familiar formula

$$\frac{\sin \pi z}{\pi z} = \prod_{n=1}^{\infty} \left(1 - \frac{z^2}{n^2}\right).$$

Example 2. The **gamma function** $\Gamma(z)$ is defined by the formula

$$\frac{1}{\Gamma(z)} = ze^{\gamma z} \prod_{n=1}^{\infty} \left(1 + \frac{z}{n}\right) e^{-z/n},$$

where γ is Euler's constant,

$$\gamma = \lim_{n\to\infty} \left(1 + \frac{1}{2} + \frac{1}{3} + \cdots + \frac{1}{n} - \log n\right).$$

It is clear that $\Gamma(z)$ is meromorphic in the entire plane and has simple poles at $z = 0, -1, -2, \ldots$. The constant γ appearing in the exponent is chosen so that $\Gamma(1) = 1$. To see this, simply observe that

$$\begin{aligned}\prod_{n=1}^{\infty} \left(1 + \frac{1}{n}\right) e^{-1/n} &= \lim_{N\to\infty} \prod_{n=1}^{N} \left(1 + \frac{1}{n}\right) e^{-1/n} \\ &= \lim_{N\to\infty} (N+1) e^{-\sum_{n=1}^{N} 1/n} = e^{-\gamma}.\end{aligned}$$

Let us put

$$P(z) = \prod_{n=1}^{\infty} \left(1 + \frac{z}{n}\right) e^{-z/n}.$$

Then $P(z-1)$ is an entire function of order 1 and has (simple) zeros at $z = 0, -1, -2, \ldots$. By Hadamard's theorem, we can write

$$P(z-1) = ze^{Az+B}P(z).$$

The values of A and B are easily determined. Taking the logarithmic derivative of both sides, we obtain

$$\sum_{n=1}^{\infty}\left(\frac{1}{z-1+n}-\frac{1}{n}\right) = \frac{1}{z}+A+\sum_{n=1}^{\infty}\left(\frac{1}{z+n}-\frac{1}{n}\right),$$

which reduces to $A = 0$. Setting $z = 1$, we find

$$1 = P(0) = e^{B}P(1) = e^{B}e^{-\gamma},$$

so that $e^{B} = e^{\gamma}$, and hence

$$P(z-1) = ze^{\gamma}P(z).$$

It follows readily from this that $\Gamma(z)$ satisfies the difference equation

$$\Gamma(z+1) = z\Gamma(z).$$

Since $\Gamma(1) = 1$, repeated application of the formula above shows that

$$\Gamma(n) = (n-1)!$$

for every positive integer n. Thus, $\Gamma(z)$ is an extension of the factorial function to nonintegral values of the argument.

From the relation

$$P(z)P(-z) = \frac{\sin \pi z}{\pi z},$$

we derive the important identity

$$\Gamma(z)\Gamma(1-z) = \frac{\pi}{\sin \pi z}.$$

PROBLEMS

1. Use Hadamard's theorem to determine the canonical factorization of $\cos\sqrt{z}$.

2. Show that

$$\cos\frac{\pi z}{4} - \sin\frac{\pi z}{4} = (1-z)\left(1+\frac{z}{3}\right)\left(1-\frac{z}{5}\right)\cdots.$$

3. Let $f(z)$ be an entire function of finite order ρ, with zeros $z_1, z_2, z_3, \ldots$ other than $z = 0$. Let p be a nonnegative integer such that

$$\sum_{n=1}^{\infty} 1/|z_n|^{p+1} < \infty,$$

and write

$$f(z) = z^m e^{g(z)} \prod_{n=1}^{\infty} E\left(\frac{z}{z_n}, p\right).$$

(Note that the product is *not necessarily* a canonical product.) Does it still follow that $g(z)$ is a polynomial?

4. Prove: if $f(z)$ is an entire function of finite order ρ, but not identically zero, then for each $\varepsilon > 0$,

$$|f(z)| > e^{-r^{\rho+\varepsilon}}$$

on circles $|z| = r$ of arbitrarily large radius.

5. Show that $\Gamma\left(\frac{1}{2}\right) = \sqrt{\pi}$.

6. Show that

$$\Gamma\left(n + \frac{1}{2}\right) = \frac{1 \cdot 3 \cdot 5 \cdots (2n-1)}{2^n}\sqrt{\pi} \quad (n = 1, 2, 3, \ldots).$$

7. Show that

$$\prod_{n=1}^{\infty}\left(1 - \frac{1}{2n}\right) e^{1/2n} = \sqrt{e^{\gamma}/\pi}.$$

(*Hint*: Put $f(z) = \Gamma'(z)/\Gamma(z)$. Show that

$$f'(z) + f'\left(z + \tfrac{1}{2}\right) = 2f'(2z),$$

and then integrate.)

8. Prove that

$$\frac{d}{dz}\left(\frac{\Gamma'(z)}{\Gamma(z)}\right) = \sum_{n=0}^{\infty} \frac{1}{(z+n)^2}.$$

9. Establish **Legendre's duplication formula**,

$$\sqrt{\pi}\,\Gamma(2z) = 2^{2z-1}\Gamma(z)\Gamma\left(z + \tfrac{1}{2}\right).$$

10. Show that $1/\Gamma(z)$ is not of exponential type. (*Hint*: Use Stirling's formula to show that $\log M(r) \sim r \log r$ as $r \to \infty$.)

11. Prove **Laguerre's theorem**: if $f(z)$ is a nonconstant entire function, real for real z, of order less than 2, and with real zeros only, then (1) the zeros of $f'(z)$ are also real and (2) $f'(z)$ vanishes once and only once between successive zeros of $f(z)$. (*Hint*: For the first assertion, show that

$$\operatorname{Im}\left(\frac{f'(z)}{f(z)}\right) = 0 \quad \text{only when} \quad \operatorname{Im} z = 0.$$

For the second, show that

$$\frac{d}{dz}\left(\frac{f'(z)}{f(z)}\right) < 0 \quad \text{when} \quad \operatorname{Im} z = 0.)$$

PART TWO. RESTRICTIONS ALONG A LINE

For the remainder of this chapter we shall be concerned mainly with the class of entire functions of exponential type. Our primary aim is to characterize those functions of this class that belong to L^2 along the real axis. These functions play an important role in both the theory and applications of entire functions.

If an entire function of exponential type is known to satisfy additional growth conditions along a line, then more can be asserted about its growth in general and hence about the distribution of its zeros. We shall show, for example, that boundedness on one line carries with it boundedness on every parallel line; a similar assertion is true for functions belonging to L^p on a given line. These results are in marked contrast with the behavior of entire functions in general, about which such assertions are not universally true.

We begin by establishing a far-reaching generalization of the maximum modulus principle.

1 THE "PHRAGMÉN–LINDELÖF" METHOD

Theorems that come under this collective heading establish the boundedness of an analytic function inside an infinite region, given that the function is known to be bounded on the boundary and not of too rapid growth inside. The fundamental result is that in which the unbounded region is an infinite sector.

Theorem 10 (Phragmén–Lindelöf). *Let $f(z)$ be continuous on a closed sector of opening π/α and analytic in the open sector. Suppose that on the bounding rays of the sector,*

$$|f(z)| \leqq M,$$

and that for some $\beta < \alpha$,

$$|f(z)| \leqq e^{r^{\beta}}$$

whenever z lies inside the sector and $|z| = r$ is sufficiently large. Then $|f(z)| \leqq M$ throughout the sector.

Proof. We can suppose without loss of generality that the given sector is symmetric with respect to the positive real axis and has its vertex at the origin. We introduce the auxiliary function

$$g(z) = e^{-\varepsilon z^{\gamma}} f(z),$$

where $\beta < \gamma < \alpha$ and $\varepsilon > 0$. Here z^{γ} denotes that single-valued analytic branch of the multiple-valued function $z^{\gamma} = \exp(\gamma \log z)$ that takes positive values for positive real z. Setting $z = re^{i\theta}$, we have

$$|g(z)| = e^{-\varepsilon r^{\gamma} \cos \gamma\theta} |f(z)|.$$

On the bounding rays of the sector, $\cos \gamma\theta$ is positive, and hence

$$|g(z)| \leqq |f(z)| \leqq M.$$

On the arc $|\theta| \leqq \pi/2\alpha$ of the circle $|z| = r$,

$$|g(z)| \leqq e^{-\varepsilon r^{\gamma} \cos \gamma\pi/2\alpha} |f(z)| < e^{r^{\beta} - \varepsilon r^{\gamma} \cos \gamma\pi/2\alpha}$$

as soon as r is sufficiently large. But this last expression approaches zero as $r \to \infty$, so that, if r is sufficiently large, $|g(z)| \leqq M$ on this arc also. By the maximum modulus principle, $|g(z)| \leqq M$ throughout that part of the sector for which $|z| \leqq r$ and hence throughout the entire sector since r can be made arbitrarily large. Therefore, we conclude that

$$|f(z)| \leqq Me^{\varepsilon |z|^{\gamma}}$$

everywhere within the sector. Since ε was arbitrary, the result follows. ∎

The theorem is sharp in the sense that the conclusion is no longer valid when $\beta = \alpha$. Indeed, we need only consider the function

$$f(z) = e^{z^{\alpha}} \quad \text{for} \quad |\arg z| \leqq \pi/2\alpha;$$

$f(z)$ is bounded on the rays of the sector but certainly not in the interior.

Corollary. *An entire function of order less than one that is bounded on a line must reduce to a constant.*

We can now prove that an entire function of exponential type that is bounded on a line must be bounded on every parallel line. There is no loss of generality in supposing that the given line is the real axis.

Theorem 11. *Let $f(z)$ be an entire function such that*

$$|f(z)| \leqq Ae^{B|z|}$$

for all values of z and

$$|f(x)| \leqq M$$

for all real values of x. Then

$$|f(x+iy)| \leqq Me^{B|y|}.$$

Proof. Suppose first that $y > 0$. Let ε be an arbitrary positive number and put

$$g(z) = e^{i(B+\varepsilon)z} f(z).$$

Then

$$|g(x)| = |f(x)| \leqq M$$

for all real x, while

$$g(iy) \to 0 \quad \text{as} \quad y \to \infty.$$

Let N denote the maximum value of $|g(z)|$ on the nonnegative imaginary axis. The Phragmén–Lindelöf theorem, applied separately to the first and second quadrants, shows that

$$|g(z)| \leqq \max(N, M)$$

throughout the upper half-plane. A simple application of the maximum modulus principle then shows that $N \leqq M$, and hence

$$|g(z)| \leqq M \quad \text{whenever} \quad \operatorname{Im} z > 0.$$

Therefore,

$$|f(z)| \leqq e^{(B+\varepsilon)y}|g(z)| \leqq Me^{(B+\varepsilon)y}$$

throughout the upper half-plane, and the result follows by letting $\varepsilon \to 0$.

The case in which $y < 0$ can be reduced to the first case by considering $f(-z)$. ∎

We note that the theorem is sharp in the sense that neither B nor M can be replaced by any smaller constant in the conclusion—simply consider $f(z) = \sin z$. We note also that the conclusion holds under the apparently weaker assumption that $f(z)$ is an entire function of exponential type at most B. For in this case we have

$$|f(z)| \leqq A(\varepsilon)e^{(B+\varepsilon)|z|}$$

for each positive number ε and all values of z, and consequently

$$|f(x+iy)| \leqq Me^{(B+\varepsilon)|y|}.$$

Now let $\varepsilon \to 0$.

Theorem 11 shows that an entire function $f(z)$ of exponential type that is bounded on the real axis is in fact "uniformly bounded" in every horizontal strip. If, in addition, we know that $f(x) \to 0$ as $|x| \to \infty$, then more can be asserted.

Theorem 12. *If $f(z)$ is an entire function of exponential type and if*

$$f(x) \to 0 \quad \text{as} \quad |x| \to \infty,$$

then

$$f(x+iy) \to 0 \quad \text{as} \quad |x| \to \infty$$

uniformly in every horizontal strip.

Proof. Since $f(z)$ is uniformly bounded in every horizontal strip, the result is an immediate consequence of Montel's theorem (see Titchmarsh [1939, p. 170]). ■

The Fourier–Stieltjes integrals

$$\int_{-\tau}^{\tau} e^{izt}\, d\omega(t),$$

where $\omega(t)$ is a function of bounded variation on the interval $[-\tau, \tau]$, provide a large class of examples of entire functions of exponential type (no larger than τ) that are bounded on the real axis. Special cases include the "almost periodic" exponential sums

$$\sum_n c_n e^{i\lambda_n z},$$

with $-\tau \leqq \lambda_n \leqq \tau$ and $\sum |c_n| < \infty$, as well as the integrals

$$\int_{-\tau}^{\tau} e^{izt} f(t)\, dt,$$

with $\int_{-\tau}^{\tau} |f(t)|\, dt < \infty$.

The totality $\boldsymbol{B}_\tau$ of all entire functions $f(z)$ of exponential type at most τ that are bounded on the real axis, together with the norm

$$\|f\| = \sup_{-\infty<x<\infty} |f(x)|,$$

constitutes one of the important classical Banach spaces of entire functions. That $\boldsymbol{B}_\tau$ is a vector space (under pointwise addition and scalar multiplication) is clear; that it is a Banach space is an easy consequence of Theorem 11.

The space $\boldsymbol{B}_\tau$ was first considered by Bernstein. The following remarkable theorem is due to him: if $f(z) \in \boldsymbol{B}_\tau$, then

$$\|f'\| \leqq \tau \|f\|;$$

moreover, equality can hold only for functions of the form

$$f(z) = \alpha e^{i\tau z} + \beta e^{-i\tau z},$$

where α and β are constants. **Bernstein's inequality,** as the aforementioned result is known, plays an important role in the theory of approximation of continuous functions (see Akhiezer [1956]; see also Problems 12 and 13 and Example 2 of Section 4 for a proof of Bernstein's inequality).

PROBLEMS

1. Let $f(z)$ be continuous in the closed square $[-1, 1] \times [-1, 1]$ and analytic in the interior. Suppose that on each bounding edge S_i $(i = 1, 2, 3, 4)$ we have

$$|f(z)| \leqq M_i \quad \text{whenever} \quad z \in S_i.$$

Prove that $|f(0)| \leqq (M_1 M_2 M_3 M_4)^{1/4}$.

2. Prove that the conclusion of Theorem 10 still holds if we are only given

$$f(z) = O(e^{\varepsilon r^\alpha})$$

for each $\varepsilon > 0$, uniformly in the sector. (*Hint*: Consider the function $e^{-\varepsilon z^\alpha} f(z)$.)

3. Prove that an entire function of exponential type zero that is bounded on a line must reduce to a constant.

4. Prove that an entire function of order $\rho < \frac{1}{2}$ that is bounded on a ray must reduce to a constant.

5. Prove that an entire function of exponential type that is bounded on two nonparallel lines must reduce to a constant.

6. Let $f(z)$ be continuous in a closed sector and analytic and bounded in its interior. Prove that if $f(z) \to a$ as $z \to \infty$ along the bounding rays of the sector, then $f(z) \to a$ uniformly in the sector. (*Hint*: Suppose first that $|\arg z| \leqq \alpha < \pi/2$ and consider the function $(z/(z + \lambda))f(z)$, with $\lambda > 0$.)

7. Let $f(z)$ be continuous in a closed sector and analytic and bounded in its interior. Suppose that $f(z) \to a$ as $z \to \infty$ along one of the bounding rays, while $f(z) \to b$ as $z \to \infty$ along the other. Prove that $a = b$. (*Hint*: Consider the function $(f(z) - \frac{1}{2}(a+b))^2$.)

8. Let $f(z)$ be analytic in an open connected set Ω. Suppose that for each boundary point of Ω and each $\varepsilon > 0$ there exists a neighborhood such that at every point z of Ω in this neighborhood one has

$$|f(z)| < M + \varepsilon.$$

Prove that $|f(z)| \leqq M$ throughout Ω (in fact, unless $f(z)$ is a constant, $|f(z)| < M$).

9. Let $f(z)$ be analytic and bounded in the right half-plane $\operatorname{Re} z > 0$. Prove that if $z_1, z_2, z_3, \ldots$ are the zeros of $f(z)$ in this half-plane, then the series

$$\sum_{n=1}^{\infty} \operatorname{Re}\left(\frac{1}{z_n}\right)$$

is convergent. (*Hint*: Prove that if $|f(z)| \leqq M$ whenever $\operatorname{Re} z > 0$, then the stronger inequality

$$|f(z)| \leqq \left|\frac{z_1 - z}{\overline{z}_1 + z} \cdots \frac{z_n - z}{\overline{z}_n + z}\right| M$$

holds for $\operatorname{Re} z > 0$.)

10. Prove that if $f(z)$ is an entire function of exponential type, bounded on the real axis, and if $z_1, z_2, z_3, \ldots$ are its zeros other than $z = 0$, then

$$\sum_{n=1}^{\infty} \left|\operatorname{Im} \frac{1}{z_n}\right| < \infty.$$

(*Hint*: Use the result of Problem 9.)

11. Use Cauchy's integral formula to show that if $f(z)$ is an entire function of exponential type that is bounded on the real axis, then $f'(z)$ is also bounded on the real axis.

12. Prove Bernstein's inequality for functions of the form

$$f(z) = \int_{-\pi}^{\pi} e^{izt}\, d\omega(t),$$

where $\omega(t)$ is of bounded variation on the interval $[-\pi, \pi]$. (*Hint*: Use the relations

$$f'(z) = \int_{-\pi}^{\pi} ite^{izt}\, d\omega(t)$$

and

$$ite^{-it/2} = \frac{4}{\pi} \sum_{n=-\infty}^{\infty} \frac{(-1)^n}{(2n+1)^2} e^{int} \quad \text{for} \quad |t| \leqq \pi,$$

to show that

$$f'(x) = \frac{4}{\pi} \sum_{n=-\infty}^{\infty} \frac{(-1)^n}{(2n+1)^2} f\left(x + n + \frac{1}{2}\right).)$$

13. **(Bernstein)** If $f(t) = \sum_{n=-N}^{N} c_n e^{int}$ is a trigonometric polynomial of degree N and if M is the maximum value of $|f(t)|$ (t real), then

$$|f'(t)| \leqq MN$$

for all real values of t.

2 CARLEMAN'S FORMULA

Jensen's formula provides a simple relation between the modulus of an analytic function on a circle and the distribution of its zeros inside the circle. If additional information is available on the growth of the function along a line, then Carleman's formula is of fundamental importance.

Theorem 13 (Carleman). *Let $f(z)$ be analytic for* $\operatorname{Im} z \geqq 0$ *and let* $z_k = r_k e^{i\theta_k}$ $(k = 1, \ldots, n)$ *be its zeros in the region*

$$\Omega = \{z\colon \operatorname{Im} z \geqq 0 \ \text{ and } \ 1 \leqq |z| \leqq R\}.$$

Then

$$\sum_{k=1}^{n} \left(\frac{1}{r_k} - \frac{r_k}{R^2}\right) \sin\theta_k = \frac{1}{\pi R} \int_0^{\pi} \log|f(Re^{i\theta})| \sin\theta \, d\theta$$

$$+ \frac{1}{2\pi} \int_1^R \left(\frac{1}{x^2} - \frac{1}{R^2}\right) \log|f(x) f(-x)| \, dx + A(R),$$

where $A(R)$ is a bounded function of R.

Proof. Suppose to begin with that $f(z)$ has no zeros lying on $\partial\Omega$ (the boundary of Ω). We consider the contour integral

$$I = \frac{1}{2\pi i} \int_{\partial\Omega} \left(\frac{1}{R^2} - \frac{1}{z^2}\right) \log f(z) \, dz,$$

where $\partial\Omega$ is assumed to be *positively* oriented and the integration begins at $z = 1$ with a fixed determination of the logarithm.

The method of proof is to evaluate I in two different ways and then equate the results. First

$$I = \frac{1}{2\pi i} \int_{\partial\Omega} \frac{d}{dz} \left(\left(\frac{z}{R^2} + \frac{1}{z}\right) \log f(z)\right) dz - \frac{1}{2\pi i} \int_{\partial\Omega} \frac{f'(z)}{f(z)} \left(\frac{z}{R^2} + \frac{1}{z}\right) dz.$$

As we make one complete passage over the contour $\partial\Omega$, $\log f(z)$ increases by $2\pi in$ (recall that n is the number of zeros of $f(z)$ inside $\partial\Omega$), so that the first integral on the right-hand side is equal to

$$n\left(\frac{1}{R^2}+1\right).$$

The second integral is easily evaluated by the residue theorem—this gives

$$I = n\left(\frac{1}{R^2}+1\right)+\sum_{k=1}^{n}\left(\frac{z_k}{R^2}+\frac{1}{z_k}\right).$$

Next, we evaluate I by integrating separately over each component of $\partial\Omega$. On the positive real axis, we obtain

$$\frac{1}{2\pi i}\int_1^R\left(\frac{1}{R^2}-\frac{1}{x^2}\right)\log f(x)\,dx,$$

and on the negative real axis,

$$\frac{1}{2\pi i}\int_{-R}^{-1}\left(\frac{1}{R^2}-\frac{1}{x^2}\right)\log f(x)\,dx;$$

on the large semicircle, $z = Re^{i\theta}$, and we obtain

$$\frac{i}{\pi R}\int_0^{\pi}\log(f(Re^{i\theta}))\sin\theta\,d\theta;$$

and, finally, on the small semicircle, we obtain a *bounded* function of R. The result now follows by equating the imaginary parts of both values of I.

The restriction that $f(z)$ be free of zeros on the contour can be eliminated by a simple continuity argument. The details are left to the reader. ∎

An entire function $f(z)$ of exponential type $\tau > 0$ that is bounded on the real axis is in many ways similar to $\sin\tau z$. It is clear, for example, that $f(z)$ must have an *infinite* number of zeros. Reason: By Hadamard's factorization theorem,

$$f(z) = z^m e^{az+b}\prod_{n=1}^{\infty}\left(1-\frac{z}{z_n}\right)e^{z/z_n};$$

if the product were finite, then $f(z)$ could not remain bounded on the real axis without reducing to a constant.

It is a simple consequence of Carleman's formula that the zeros of $f(z)$ must cluster about the real axis. Specifically, we have the following result (cf. Problem 10, Section 1).

Theorem 14. *If $f(z)$ is an entire function of exponential type, bounded on the real axis, and if $z_n = r_n e^{i\theta_n}$ $(n = 1, 2, 3, \ldots)$ are the zeros of $f(z)$ other than $z = 0$, then the series*

$$\sum_{n=1}^{\infty} \frac{\sin \theta_n}{r_n}$$

is absolutely convergent.

Proof. Since $|f(x)|$ is bounded on the real axis, the integrals

$$\int_1^R \left(\frac{1}{x^2} - \frac{1}{R^2} \right) \log |f(x) f(-x)| \, dx$$

have a finite upper bound for all values of $R > 1$. In addition, since $f(z)$ is of exponential type, there is a constant K such that

$$\log |f(Re^{i\theta})| \leqq KR,$$

as soon as R is sufficiently large, so that

$$\frac{1}{\pi R} \int_0^{\pi} \log |f(Re^{i\theta})| \sin \theta \, d\theta \leqq \frac{2K}{\pi}.$$

Applying Carleman's formula to $f(z)$ in the upper and lower half-planes and then adding the result, we conclude that

$$\sum_{r_n < R} \left(1 - \frac{r_n^2}{R^2} \right) \frac{|\sin \theta_n|}{r_n} < A$$

for some constant A and all large values of R. But this implies that

$$\sum_{r_n < R/2} \left(1 - \frac{1}{4} \right) \frac{|\sin \theta_n|}{r_n} < A,$$

and the result follows by letting $R \to \infty$. ■

Let us write the conclusion of Theorem 14 as follows:

$$\sum_{n=1}^{\infty} \left| \operatorname{Im} \frac{1}{z_n} \right| < \infty.$$

Then, for each positive number ε, the zeros z_{n_k} of $f(z)$ that lie outside the sectors $|\arg z| < \varepsilon$ and $|\arg z - \pi| < \varepsilon$ must satisfy

$$\sum_{k=1}^{\infty} \frac{1}{|z_{n_k}|} < \infty.$$

Now, it can be shown that if $f(z)$ is of exponential type $\tau > 0$ and bounded on the real axis, then the set consisting of *all* its zeros has a "density" equal to $2\tau/\pi$, i.e.,

$$\lim_{r\to\infty} \frac{n(r)}{r} = \frac{2\tau}{\pi}$$

(this is hard; see, for example, Levinson [1940, Chap. III]). Accordingly,

$$\sum_{n=1}^{\infty} \frac{1}{|z_n|} = \infty,$$

and hence "most" of the zeros of $f(z)$ lie "near" the real axis.

Theorem 14 may be reformulated as a *uniqueness* theorem: an entire function of exponential type that is bounded on the real axis is completely determined by its values on any set for which $\sum |\mathrm{Im}\, 1/z_n| = \infty$. As an application, we shall prove the following completeness theorem for systems of complex exponentials in $C[a, b]$.

Theorem 15. *If $\{\lambda_n\}$ is a sequence of distinct complex numbers for which $|\arg \lambda_n - \pi/2| \leqq L < \pi/2$ and if*

$$\sum_{n=1}^{\infty} \frac{1}{|\lambda_n|} = \infty,$$

then the system $\{e^{i\lambda_n t}\}$ is complete in $C[a, b]$ for every finite interval $[a, b]$.

Proof. We argue by contradiction. If the system failed to be complete in $C[a, b]$ for some interval $[a, b]$, then by virtue of the Riesz representation theorem we could find an entire function $f(z)$ of the form

$$f(z) = \int_a^b e^{izt}\, d\omega(t),$$

with $\omega(t)$ of bounded variation on $[a, b]$, such that $f(z)$ vanishes at every λ_n but does not vanish identically. According to Theorem 14, the series

$$\sum_{n=1}^{\infty} \mathrm{Im}\, \frac{1}{\lambda_n}$$

is absolutely convergent. But

$$\sum_{n=1}^{\infty} \left| \mathrm{Im}\, \frac{1}{\lambda_n} \right| \geqq \cos L \sum_{n=1}^{\infty} \frac{1}{|\lambda_n|} = \infty,$$

and the contradiction proves the theorem. ■

By choosing the λ_n to be purely imaginary, we obtain at once the following version of Müntz's theorem on polynomial approximation.

Corollary (Müntz). *If* $\{\lambda_1, \lambda_2, \lambda_3, \ldots\}$ *is an increasing sequence of positive real numbers such that*

$$\sum_{n=1}^{\infty} \frac{1}{\lambda_n} = \infty,$$

then the set of powers

$$\{t^{\lambda_1}, t^{\lambda_2}, t^{\lambda_3}, \ldots\}$$

is complete in $C[a, b]$ *whenever* $0 < a < b$.

PROBLEMS

1. Establish the following version of Carleman's theorem: if $f(z)$ is analytic for $\operatorname{Im} z \geqq 0$, if $z_1, \ldots, z_n$ are its zeros in the region $\{z: |z| < R, \operatorname{Im} z > 0\}$, and if $f(0) = 1$, then

$$\sum_{k=1}^{n} \left(\frac{1}{r_k} - \frac{r_k}{R^2}\right) \sin\theta_k = \frac{1}{\pi R} \int_0^{\pi} \log|f(Re^{i\theta})| \sin\theta \, d\theta$$

$$+ \frac{1}{2\pi} \int_0^R \left(\frac{1}{x^2} - \frac{1}{R^2}\right) \log|f(x)f(-x)| \, dx + \frac{1}{2} \operatorname{Im} f'(0).$$

2. Show that Theorem 13 remains valid if $f(z)$ is analytic for $\operatorname{Im} z > 0$ and continuous for $\operatorname{Im} z \geqq 0$.

3. Let $f(z)$ be an entire function of exponential type and suppose that

$$\int_1^R \frac{\log|f(x)f(-x)|}{x^2} \, dx < M$$

for some constant M and all values of $R > 1$. Prove that if $z_1, z_2, z_3, \ldots$ are the zeros of $f(z)$ other than $z = 0$, then

$$\sum_{n=1}^{\infty} \left| \operatorname{Im} \frac{1}{z_n} \right| < \infty.$$

(*Hint*: Set

$$g(x) = \int_1^x \frac{\log|f(t)f(-t)|}{t^2} \, dt$$

and integrate by parts.)

4. Let $f(z)$ be analytic for $\operatorname{Re} z \geqq 0$ and suppose that for some positive number a,

$$f(z) = O(e^{-a|z|}) \quad \text{as} \quad z \to \infty,$$

uniformly in the sector. Show that $f(z)$ is identically zero. (*Hint*: Consider the function $g(z) = f(z)\sin bz$, where $0 < b < a$.)

5. Let

$$f(z) = \int_a^b e^{zt}\phi(t)\,dt,$$

where $\phi(t)$ is a real integrable function on $[a, b]$. Show that if $z_1, z_2, z_3, \ldots$ are the zeros of $f(z)$ other than $z = 0$, then

$$\sum_{n=1}^{\infty} \left|\operatorname{Re}\left(\frac{1}{z_n}\right)\right| < \infty.$$

(*Hint*: Show that $e^{az} f(z)$ is bounded for $\operatorname{Re} z \leqq 0$ and $e^{-bz} f(z)$ is bounded for $\operatorname{Re} z \geqq 0$.)

6. (a) Show that Müntz's theorem need not hold if $a < 0$.

(b) Prove that the set $\{1, t^{\lambda_1}, t^{\lambda_2}, \ldots\}$ is complete in $C[0, 1]$ whenever $\{\lambda_n\}$ is an increasing sequence of positive numbers for which $\sum 1/\lambda_n = \infty$. (*Hint*: First show that the set $\{t^{\lambda_1}, t^{\lambda_2}, t^{\lambda_3}, \ldots\}$ is complete in $L^2[0, 1]$ by considering the function

$$\int_0^1 t^z \phi(t)\,dt,$$

where $\phi \in L^2[0, 1]$ and $\operatorname{Re} z \geqq 0$. Complete the proof by observing that for $0 \leqq t \leqq 1$ and $n = 1, 2, 3, \ldots$,

$$\left|t^n - \sum c_i t^{\lambda_i}\right| = n \left|\int_0^t (x^{n-1} - \sum d_i x^{\lambda_i - 1})\,dx\right|$$

$$\leqq n \left(\int_0^1 \left|x^{n-1} - \sum d_i x^{\lambda_i - 1}\right|^2 dx\right)^{1/2}.)$$

3 INTEGRABILITY ON A LINE

In this section we shall prove that an entire function of exponential type that belongs to L^p along the real axis must also belong to L^p along every line parallel to the real axis. In fact, much more is true.

Theorem 16 (Plancherel–Pólya). *If $f(z)$ is an entire function of exponential type τ and if for some positive number p,*

$$\int_{-\infty}^{\infty} |f(x)|^p\,dx < \infty,$$

then

$$\int_{-\infty}^{\infty} |f(x+iy)|^p \, dx \leqq e^{p\tau|y|} \int_{-\infty}^{\infty} |f(x)|^p \, dx.$$

Plancherel and Pólya [1938] have given an "elementary" proof of Theorem 16 based on the Phragmén–Lindelöf method, and it is this proof that we reproduce here. The proof will require two preliminary lemmas.

Let $g(z)$ be continuous in the closed upper half-plane, $\operatorname{Im} z \geqq 0$, analytic in the interior, and suppose that $g(z)$ does not reduce to a constant. Let a and p be positive real numbers and put

$$G(z) = \int_{-a}^{a} |g(z+t)|^p \, dt.$$

It is clear that $G(z)$ is defined and continuous for $\operatorname{Im} z \geqq 0$. Since $|g(z)|^p$ is subharmonic for $\operatorname{Im} z > 0$ (see Rudin [1966, p. 329]), so is $G(z)$.

Lemma 1. *Let $g(z)$ be of exponential type in the half-plane* $\operatorname{Im} z \geqq 0$ *and suppose that the following quantities are both finite:*

$$M = \sup_{-\infty < x < \infty} G(x) \quad \text{and} \quad N = \sup_{y > 0} G(iy).$$

Then throughout this half-plane,

$$G(z) \leqq \max(M, N).$$

Proof. Since $g(z)$ is of exponential type, there exist positive numbers A and B such that

$$|g(z)| \leqq Ae^{B|z|} \quad \text{for} \quad \operatorname{Im} z \geqq 0. \tag{1}$$

For each positive number ε define the auxiliary function

$$g_\varepsilon(z) = g(z) e^{-\varepsilon(\lambda(z+a))^{3/2}}, \tag{2}$$

where $\lambda = e^{-i\pi/4}$. The exponent of e appearing in (2) has two possible determinations in the half-plane $\operatorname{Im} z > 0$; we choose the one whose real part is negative in the quarter-plane $x > -a$, $y \geqq 0$. Put

$$G_\varepsilon(z) = \int_{-a}^{a} |g_\varepsilon(z+t)|^p \, dt,$$

which is then defined and continuous in the upper half-plane $\operatorname{Im} z \geqq 0$, and subharmonic in the interior. A simple calculation involving (1) and (2) shows that in the quarter plane $x > -a$, $y \geqq 0$,

$$|g_\varepsilon(z)| \leqq Ae^{B|z| - \varepsilon\gamma|z+a|^{3/2}}, \tag{3}$$

where $\gamma = \cos 3\pi/8$, and

$$|g_\varepsilon(z)| \leqq |g(z)|.$$

Hence

$$G_\varepsilon(z) \leqq G(z) \quad \text{for} \quad x \geqq 0,\, y \geqq 0,$$

and in particular

$$G_\varepsilon(x) \leqq M \quad \text{for} \quad x \geqq 0 \quad \text{and} \quad G_\varepsilon(iy) \leqq N \quad \text{for} \quad y \geqq 0.$$

Let z_0 be a fixed but arbitrary point in the first quadrant. We shall apply the maximum principle to $G_\varepsilon(z)$ in the region $\Omega = \{z\colon \operatorname{Re} z \geqq 0, \operatorname{Im} z \geqq 0, |z| \leqq R\}$, choosing R large enough so that (i) $z_0 \in \Omega$, and (ii) the maximum value of $G_\varepsilon(z)$ on Ω is *not* attained on the circular arc $|z| = R$ (this is possible by virtue of (3)). Since $G_\varepsilon(z)$ does not reduce to a constant, the maximum value of $G_\varepsilon(z)$ on Ω must be attained on one of the coordinate axes, and hence, in particular,

$$G_\varepsilon(z_0) \leqq \max(M, N).$$

Now let $\varepsilon \to 0$. This establishes the result for the first quadrant; the proof for the second quadrant is the same. ■

Lemma 2. *In addition to the hypotheses of Lemma 1, suppose that*

$$\lim_{y\to\infty} g(x + iy) = 0 \tag{4}$$

uniformly in x, for $-a \leqq x \leqq a$. *Then*

$$G(z) \leqq M \quad \textit{whenever} \quad \operatorname{Im} z \geqq 0.$$

Proof. It is sufficient to show that $N \leqq M$. By virtue of (4), we see that the function $G(iy)$ approaches zero as $y \to \infty$, and so must attain its least upper bound N for some finite value of y, say $y = y_0$. If $y_0 = 0$, then

$$N = G(iy_0) = G(0) \leqq M.$$

If, on the other hand, $y_0 > 0$, then the maximum principle shows that the least upper bound of $G(z)$ in the half-plane $\operatorname{Im} z \geqq 0$ cannot be attained at the interior point $z = iy_0$. Therefore, by Lemma 1,

$$N = G(iy_0) < \max(M, N),$$

and so $N < M$. ■

Theorem 16 can now be established.

Proof of Theorem 16. It is sufficient to prove the theorem when $y > 0$ and $f(z)$ is not identically zero. Let ε be a fixed positive number and consider the function

$$g(z) = f(z)e^{i(\tau+\varepsilon)z}.$$

It is a simple matter to verify that for each positive number a, the functions $g(z)$ and $G(z)$ fulfill all of the necessary hypotheses of Lemmas 1 and 2. Consequently, if $y > 0$, then by Lemma 2

$$G(iy) \leqq M < \int_{-\infty}^{\infty} |g(x)|^p \, dx.$$

This together with the definitions of $g(z)$ and $G(z)$ implies

$$e^{-p(\tau+\varepsilon)y} \int_{-a}^{a} |f(x+iy)|^p \, dx < \int_{-\infty}^{\infty} |f(x)|^p \, dx.$$

The proof is over: first let $a \to \infty$, then let $\varepsilon \to 0$. ■

An important corollary to Theorem 16 asserts that $f(x) \to 0$ as $|x| \to \infty$ whenever $f(z)$ is of exponential type and belongs to L^p along the real axis. The next theorem will prove even more.

Theorem 17. *Let $f(z)$ be an entire function of exponential type τ and suppose that for some positive number p,*

$$\int_{-\infty}^{\infty} |f(x)|^p \, dx < \infty.$$

If $\{\lambda_n\}$ is an increasing sequence of real numbers such that

$$\lambda_{n+1} - \lambda_n \geqq \varepsilon > 0,$$

then

$$\sum_n |f(\lambda_n)|^p \leqq B \int_{-\infty}^{\infty} |f(x)|^p \, dx,$$

where B is a constant that depends only on p, τ, and ε.

Proof. Since $|f|^p$ is subharmonic, we know that the inequality

$$|f(z_0)|^p \leqq \frac{1}{2\pi} \int_0^{2\pi} |f(z_0 + re^{i\theta})|^p \, d\theta \tag{5}$$

holds for *all* values of r (see Rudin [1966, p. 329]), and hence that

$$|f(z_0)|^p \leqq \frac{1}{\pi\delta^2} \iint_{|z-z_0| \leqq \delta} |f(z)|^p \, dx \, dy$$

for every z_0 and every $\delta > 0$ (multiply both sides of (5) by r and integrate between 0 and δ). Therefore,

$$\sum |f(\lambda_n)|^p \leqq \frac{1}{\pi\delta^2} \sum \iint_{|z| \leqq \delta} |f(\lambda_n + z)|^p \, dx \, dy$$

$$\leqq \frac{1}{\pi\delta^2} \sum \int_{-\delta}^{\delta} \int_{-\delta}^{\delta} |f(\lambda_n + x + iy)|^p \, dx \, dy$$

$$= \frac{1}{\pi\delta^2} \sum \int_{-\delta}^{\delta} \int_{\lambda_n - \delta}^{\lambda_n + \delta} |f(x + iy)|^p \, dx \, dy.$$

Take $\delta = \varepsilon/2$. Then the intervals $(\lambda_n - \delta, \lambda_n + \delta)$ are pairwise disjoint, and the last expression above is no larger than

$$\frac{1}{\pi\delta^2} \int_{-\delta}^{\delta} \int_{-\infty}^{\infty} |f(x + iy)|^p \, dx \, dy.$$

Applying Theorem 16, we conclude that

$$\sum |f(\lambda_n)|^p \leqq \frac{1}{\pi\delta^2} \int_{-\delta}^{\delta} \left(e^{p\tau|y|} \int_{-\infty}^{\infty} |f(x)|^p \, dx \right) dy$$

$$= B \int_{-\infty}^{\infty} |f(x)|^p \, dx,$$

where $B = B(p, \tau, \varepsilon)$. ■

Remark. A sequence $\{\lambda_n\}$ of real or complex numbers is said to be *separated* if for some positive number ε,

$$|\lambda_n - \lambda_m| \geqq \varepsilon \quad \text{whenever} \quad n \neq m.$$

The theorem is easily modified to allow for complex λ_n's, provided they are separated and

$$\sup_n |\operatorname{Im} \lambda_n| < \infty.$$

The details are left to the reader.

Corollary. *If $f(z)$ is an entire function of exponential type and if*

$$\int_{-\infty}^{\infty} |f(x)|^p \, dx < \infty$$

for some positive number p, then $f(x) \to 0$ as $|x| \to \infty$.

PROBLEMS

1. Give an example of an entire function that belongs to L^1 along the real axis but does not belong to L^1 along any other line.

2. Prove that if $f(z)$ is an entire function of exponential type zero and $\int_{-\infty}^{\infty} |f(x)|^p \, dx < \infty$ for some positive value of p, then $f(z)$ is identically zero.

3. Verify the *Remark* following Theorem 17.

4. Prove: if $f(z)$ is an entire function of exponential type τ and

$$\int_{-\infty}^{\infty} |f(x)|^p \, dx < \infty$$

for some positive value of p, then

$$|f(x+iy)| \leqq Ce^{\tau|y|} \left(\int_{-\infty}^{\infty} |f(x)|^p \, dx \right)^{1/p},$$

where C is a constant depending only on p and τ. (*Hint*: For every $\delta > 0$,

$$|f(x+iy)|^p \leqq \frac{1}{\pi\delta^2} \int_{y-\delta}^{y+\delta} \int_{-\infty}^{\infty} |f(s+it)|^p \, ds \, dt.)$$

5. Show that the space $\boldsymbol{E}_\tau^p (p \geqq 1, \tau > 0)$ consisting of all entire functions $f(z)$ of exponential type at most τ for which

$$||f||_p = \left(\int_{-\infty}^{\infty} |f(x)|^p \, dx \right)^{1/p} < \infty$$

is a Banach space.

6. Show that if $\phi(t) \in L^2[-\tau, \tau]$, then

$$f(z) = \int_{-\tau}^{\tau} \phi(t) e^{izt} \, dt$$

belongs to $\boldsymbol{E}_\tau^2$ (see Problem 5).

7. **(Plancherel–Pólya)** If $f(z)$ is an entire function of exponential type τ and $\int_{-\infty}^{\infty} |f(x)|^p \, dx < \infty$ for some positive value of p, then $f'(z)$ has the same properties. Moreover, $\int_{-\infty}^{\infty} |f'(x)|^p \, dx \leqq A \int_{-\infty}^{\infty} |f(x)|^p \, dx$, where A depends only on p and τ. (*Hint*: If $\delta > 0$, then

$$|f'(z_0)|^p \leqq C \int_{-\delta}^{\delta} \int_{-\delta}^{\delta} |f(z_0 + x + iy)|^p \, dx \, dy,$$

where C depends only on p and δ. To prove this, apply inequality (5) to the function

$$g(z) = \frac{f(z_0 + z) - f(z_0)}{z}$$

inside a circle of radius r centered at $z = 0$ and obtain

$$|f'(z_0)|^p \leqq \frac{2^p}{\pi r^p} \int_0^{2\pi} |f(z_0 + re^{i\theta})|^p \, d\theta.$$

Now multiply both sides by r^{p+1} and integrate with respect to r between 0 and δ. (With Theorem 18 of the next section, the problem becomes trivial for $p = 2$; see the remarks preceding Theorem 19.))

4 THE PALEY–WIENER THEOREM

It is a relatively simple matter to exhibit a large class of examples of entire functions $f(z)$ of exponential type for which

$$\int_{-\infty}^{\infty} |f(x)|^2 \, dx < \infty.$$

If $\phi \in L^2[-A, A]$, then

$$f(z) = \int_{-A}^{A} \phi(t) e^{izt} \, dt$$

is such a function: a straightforward application of Morera's theorem shows that $f(z)$ is entire; the estimate

$$|f(z)| \leqq \int_{-A}^{A} |\phi(t)| e^{-yt} \, dt \leqq e^{A|y|} \int_{-A}^{A} |\phi(t)| \, dt$$

shows that $f(z)$ is of exponential type at most A; and Plancherel's theorem shows that $\int_{-\infty}^{\infty} |f(x)|^2 \, dx = 2\pi \int_{-A}^{A} |\phi(t)|^2 \, dt < \infty$.

It is a remarkable fact that there are no other examples — every entire function of exponential type that belongs to L^2 along the real axis is obtained in this way. This is the content of the celebrated "Paley–Wiener Theorem".

Theorem 18 (Paley–Wiener). *Let $f(z)$ be an entire function such that*

$$|f(z)| \leqq Ce^{A|z|} \tag{1}$$

for positive constants A and C and all values of z, and

$$\int_{-\infty}^{\infty} |f(x)|^2 \, dx < \infty.$$

Then there exists a function ϕ in $L^2[-A, A]$ such that

$$f(z) = \int_{-A}^{A} \phi(t) e^{izt} \, dt. \tag{2}$$

Proof (Boas). Let $\phi(t)$ be the Fourier transform of $f(x)$, i.e.,

$$\phi(t) = \frac{1}{2\pi}\int_{-\infty}^{\infty} f(x)e^{-ixt}\,dx,$$

where the integral is to be interpreted as a *limit in the mean* in the L^2 sense. Then $\phi \in L^2(-\infty, \infty)$ and, by the Fourier inversion formula,

$$f(x) = \int_{-\infty}^{\infty} \phi(t)e^{ixt}\,dt.$$

Accordingly, we need only show that $\phi(t)$ vanishes almost everywhere outside the interval $[-A, A]$. This will establish (2) for all real values of z and hence for all complex values as well since both sides of (2) are entire.

Let T be a positive number. We shall consider the contour integral

$$I = \int_{\gamma} f(z)e^{-izt}\,dz,$$

where t is fixed and γ consists of the upper three sides of the rectangle $[-T, T] \times [0, T]$. Since the integrand is an entire function of z for each value of t, we have

$$I = -\int_{-T}^{T} f(x)e^{-ixt}\,dx$$

by Cauchy's integral theorem. We are going to show that $I \to 0$ as $T \to \infty$ whenever $|t| > A$. Since

$$\int_{-T}^{T} f(x)e^{-ixt}\,dt \to \phi(t) \quad \textit{in the mean} \quad (\text{as } T \to \infty),$$

it will follow that $\phi(t) = 0$ almost everywhere outside $[-A, A]$.

Suppose first that $t < -A$. Integrating over each of the three sides of γ, we readily obtain

$$|I| \leqq \int_0^T e^{ty}|f(T + iy)|\,dy + e^{tT}\int_{-T}^{T} |f(x + iT)|\,dx$$
$$+ \int_0^T e^{ty}|f(-T + iy)|\,dy \equiv I_1 + I_2 + I_3.$$

We shall estimate the size of each of these three integrals. Since $\int_{-\infty}^{\infty} |f(x)|^2\,dx$ is finite, it follows that $f(x) \to 0$ as $|x| \to \infty$ (by the corollary to Theorem 17), and hence, in particular, that $|f(x)|$ has a finite upper bound M on the real axis. By Theorem 11,

$$|f(x + iT)| \leqq Me^{AT}$$

for all real values of x, and hence

$$I_2 \leqq 2TMe^{(t+A)T}.$$

Since $t < -A$, the exponent above is negative, and we conclude that $I_2 \to 0$ as $T \to \infty$.

Turning next to I_1 (I_3 is treated in the same way), we write

$$I_1 = \left(\int_0^R + \int_R^T\right) e^{ty}|f(T+iy)|\,dy \quad (R > 0).$$

Theorem 12 shows that for each fixed R,

$$f(T+iy) \to 0 \quad \text{as} \quad T \to \infty,$$

uniformly in y, $0 \leqq y \leqq R$. Consequently,

$$\int_0^R e^{ty}|f(T+iy)|\,dy \to 0 \quad \text{as} \quad T \to \infty.$$

It remains only to show that $\int_R^T e^{ty}|f(T+iy)|\,dy$ can be made arbitrarily small by choosing R and T sufficiently large. Again appealing to Theorem 11, we have

$$\int_R^T e^{ty}|f(T+iy)|\,dy \leqq M\int_R^T e^{(t+A)y}\,dy = \frac{M}{t+A}(e^{(t+A)T} - e^{(t+A)R}),$$

and since $t + A$ is negative, the last term approaches zero as R and T approach infinity.

We have thus shown that $I \to 0$ as $T \to \infty$ whenever $t < -A$. The case $t > A$ is treated similarly, with γ in the lower half-plane. The details, which are obvious, are left to the reader. ■

It is important to note that condition (1) may be replaced by the seemingly weaker assumption that $f(z)$ is entire of exponential type at most A. For, if this is the case, then

$$|f(z)| \leqq C(\varepsilon)e^{(A+\varepsilon)|z|}$$

for every positive number ε, and the proof of Theorem 18 shows that the Fourier transform of $f(x)$ must vanish almost everywhere outside the interval $[-A-\varepsilon, A+\varepsilon]$, and so almost everywhere outside $[-A, A]$.

The following useful corollary is now readily established.

Corollary. *If $f(z)$ is an entire function of exponential type A, belonging to L^2 on the real axis, then*

$$|f(z)|e^{-A|y|} \to 0 \quad \textit{as} \quad |z| \to \infty.$$

The Paley–Wiener theorem has a wide range of important and varied applications, several of which are illustrated by the following examples. We begin with an integral representation for entire functions of exponential type that are bounded along the real axis.

Example 1. If $f(z) \in \boldsymbol{B}_\tau$, then

$$f(z) = f(0) + z \int_{-\tau}^{\tau} \phi(t) e^{izt}\, dt,$$

where $\phi \in L^2[-\tau, \tau]$.

Proof. It is clear that the function

$$\frac{f(z) - f(0)}{z}$$

is entire of exponential type at most τ and belongs to L^2 along the real axis. Now apply Theorem 18. ∎

Example 2 (Bernstein's inequality). If $f(z) \in \boldsymbol{B}_\tau$ and $|f(x)| \leqq M$ for all real values of x, then

$$|f'(x)| \leqq \tau M.$$

Proof. The inequality is easily established for functions of the form

$$\int_{-\tau}^{\tau} e^{izt}\, d\omega(t), \tag{3}$$

where $\omega(t)$ is of bounded variation on the interval $[-\tau, \tau]$ (see Problem 12, Section 1).

Suppose now that $f(z)$ is an arbitrary function belonging to $\boldsymbol{B}_\tau$. If we put

$$g_\varepsilon(z) = f(z) \frac{\sin \varepsilon z}{\varepsilon z} \qquad (\varepsilon > 0),$$

then $g_\varepsilon(z)$ is entire of exponential type at most $\tau + \varepsilon$ and belongs to L^2 along the real axis. By virtue of the Paley–Wiener theorem, $g_\varepsilon(z)$ is of the form (3) with $\tau + \varepsilon$ in place of τ. Since $|g(x)| \leqq M$, it follows that

$$|g_\varepsilon'(x)| \leqq M(\tau + \varepsilon)$$

for every real x. But $g_\varepsilon'(x) \to f'(x)$ as $\varepsilon \to 0$, and the result follows. ∎

As a final application, we shall establish an interesting quadrature formula for a certain class of entire functions of exponential type.

Example 3. If $f(z)$ is an entire function of exponential type at most 2π and $\int_{-\infty}^{\infty} |f(x)|\,dx < \infty$, then

$$\sum_{n=-\infty}^{\infty} f(n) = \int_{-\infty}^{\infty} f(x)\,dx.$$

Proof. Since $f(x)$ is integrable over $(-\infty, \infty)$, so is $f'(x)$ (Problem 7, Section 3) and $f(x) \to 0$ as $|x| \to \infty$. Under these conditions, the *Poisson summation formula* is applicable: if F is the Fourier transform of f, i.e., if $F(t) = (1/2\pi)\int_{-\infty}^{\infty} f(x)e^{-ixt}\,dx$, then

$$\sum_{n=-\infty}^{\infty} f(n) = 2\pi \sum_{n=-\infty}^{\infty} F(2\pi n).$$

But $\int_{-\infty}^{\infty} |f(x)|^2\,dx < \infty$, and the Paley–Wiener theorem implies that $F(t) = 0$ for *every* t with $|t| \geqq 2\pi$ (note that $F(t)$ is continuous). Conclusion:

$$\sum_{n=-\infty}^{\infty} f(n) = 2\pi F(0) = \int_{-\infty}^{\infty} f(x)\,dx. \quad \blacksquare$$

PROBLEMS

1. Let $f(z) = \int_{-A}^{A} \phi(t)e^{izt}\,dt$, where $\phi \in L^2[-A, A]$, and suppose that ϕ does not vanish almost everywhere in any neighborhood of A. Show that $f(z)$ is of order 1 and exponential type A.
2. Establish the corollary to Theorem 18.
3. Prove the following "gap" theorem for entire functions: if $f(z)$ is an entire function of exponential type, bounded on the real axis, and if

$$f^{(\lambda_n)}(0) = 0$$

for a sequence of positive integers λ_n for which

$$\sum_{\lambda_n \text{ even}} \frac{1}{\lambda_n} = \infty \quad \text{and} \quad \sum_{\lambda_n \text{ odd}} \frac{1}{\lambda_n} = \infty,$$

then $f(z)$ must reduce to a constant. (*Hint*: Put $g(z) = (f(z) - f(0))/z$. Then $g(z) = \int_{-A}^{A} \phi(t)e^{izt}\,dt$, with ϕ in $L^2[-A, A]$. Consider the functions

$$\int_0^A t^z(\phi(t) + \phi(-t))\,dt \quad \text{and} \quad \int_0^A t^z(\phi(t) - \phi(-t))\,dt.)$$

5 THE PALEY–WIENER SPACE

The totality of all entire functions of exponential type at most π that are square integrable on the real axis will henceforth be known as the **Paley–Wiener space** and will be designated by $\boldsymbol{P}$. Clearly, $\boldsymbol{P}$ is a vector space under pointwise addition and scalar multiplication; it is also an inner product space with respect to the inner product

$$(f, g) = \int_{-\infty}^{\infty} f(x)\overline{g(x)}\, dx.$$

Since the Fourier transform is an isometry, the Paley–Wiener theorem shows that $\boldsymbol{P}$ is a separable Hilbert space, isometrically isomorphic to $L^2[-\pi, \pi]$.

The isomorphism between $\boldsymbol{P}$ and $L^2[-\pi, \pi]$ has far-reaching consequences. At present, we shall investigate some of the ways in which known properties of $L^2[-\pi, \pi]$ can be transformed easily into nontrivial assertions about $\boldsymbol{P}$. This will then serve as the basis for a more penetrating analysis of nonharmonic Fourier series in $L^2[-\pi, \pi]$.

If f belongs to $\boldsymbol{P}$ and has the representation

$$f(z) = \frac{1}{2\pi} \int_{-\pi}^{\pi} \phi(t) e^{izt}\, dt, \tag{1}$$

with ϕ in $L^2[-\pi, \pi]$, then Plancherel's theorem shows that

$$\|f\|^2 = \int_{-\infty}^{\infty} |f(x)|^2\, dx = \frac{1}{2\pi} \int_{-\pi}^{\pi} |\phi(t)|^2\, dt = \|\phi\|^2.$$

When Parseval's identity is applied to ϕ, we obtain the important formula

$$\|f\|^2 = \sum_{-\infty}^{\infty} |f(n)|^2.$$

By taking the Fourier transform of e^{int} $(n = 0, \pm 1, \pm 2, \ldots)$, we see that the set of functions $\{(\sin \pi(z-n))/\pi(z-n)\}_{-\infty}^{\infty}$ forms an orthonormal basis for $\boldsymbol{P}$. Accordingly, each function f in $\boldsymbol{P}$ has a unique expansion of the form

$$f(z) = \sum_{n=-\infty}^{\infty} c_n \frac{\sin \pi(z-n)}{\pi(z-n)},$$

with $\sum |c_n|^2 < \infty$. The convergence of the series is understood to be in the metric of $\boldsymbol{P}$. But convergence in $\boldsymbol{P}$ implies uniform convergence in each horizontal strip. This is an immediate consequence of the following useful estimate:

$$|f(x+iy)| \leqq e^{\pi|y|}\, \|f\|. \tag{2}$$

(Proof: Simply use (1) together with the fact that $||\phi|| = ||f||$.) If we now set $z = n$, it follows that $c_n = f(n)$, and we have thereby obtained the **cardinal series** for f:

$$f(z) = \sin \pi z \sum_{n=-\infty}^{\infty} (-1)^n \frac{f(n)}{\pi(z-n)}.$$

Thus, a function in $\boldsymbol{P}$ can be recaptured from its values at the integers.

The cardinal series was first introduced by Whittaker [1915]. It plays an important role in the mathematical theory of communication, where it is known as the *sampling theorem* (see Shannon [1949] and Shannon and Weaver [1964]).

The Paley–Wiener representation yields a trivial proof that $\boldsymbol{P}$ is "closed" under differentiation (cf. Problem 7, Section 3). Indeed, if $f(z)$ is given by (1), then

$$f'(z) = \frac{1}{2\pi} \int_{-\pi}^{\pi} it\phi(t)e^{izt}\,dt$$

and $t\phi(t) \in L^2[-\pi, \pi]$. Moreover, we have the following sharp estimate for the norm of $f'(z)$:

$$\|f'\| \leqq \pi\|f\|.$$

(Proof: $\|f'\| = \|t\phi(t)\| \leqq \pi\|\phi\| = \pi\|f\|$.)

Theorem 19. *The Paley–Wiener space is a functional Hilbert space of entire functions. Its reproducing kernel K is given by*

$$K(z, w) = \frac{\sin \pi(z - \overline{w})}{\pi(z - \overline{w})},$$

and the integral representation

$$f(z) = \int_{-\infty}^{\infty} f(t) \frac{\sin \pi(t - z)}{\pi(t - z)}\,dt$$

is valid for every function f belonging to the space.

Proof. Inequality (2) shows that "point-evaluations" on $\boldsymbol{P}$ are bounded linear functionals. Thus, $\boldsymbol{P}$ is a *functional* Hilbert space.

If $f(z) \in \boldsymbol{P}$, then the Paley–Wiener theorem shows that

$$f(w) = (\phi(t), e^{i\overline{w}t})$$

for some function ϕ in $L^2[-\pi, \pi]$ and all values of w. But the Fourier transform is an isomorphism between $L^2[-\pi, \pi]$ and $\boldsymbol{P}$, and hence

$$f(w) = \left(f(z), \frac{\sin \pi(z - \overline{w})}{\pi(z - \overline{w})} \right).$$

Since $K(z, w)$ is the *unique* function on $\boldsymbol{C} \times \boldsymbol{C}$ for which

$$f(w) = (f(z), K(z, w)),$$

it follows that $K(z, w) = (\sin \pi(z - \overline{w}))/\pi(z - \overline{w})$. ■

Corollary. *If $f \in \boldsymbol{P}$ and $z = x + iy$, then*

$$|f(z)|^2 \leqq \|f\|^2 \frac{\sin 2\pi y}{2\pi y}.$$

PROBLEMS

1. Show that if $f \in \boldsymbol{P}$, then

$$e^{-\pi|y|} \|f\| \leqq \left(\int_{-\infty}^{\infty} |f(x+iy)|^2 dx \right)^{1/2} \leqq e^{\pi|y|} \|f\|.$$

2. Let $f(z)$ be an entire function of exponential type at most π, and suppose that

$$f(x) = O(x^\alpha), \quad 0 < \alpha < \tfrac{1}{2}.$$

Show that

$$f(z) = f(0) + \frac{f'(0) \sin \pi z}{\pi} + z \sin \pi z \sum_{n \neq 0} \frac{(-1)^n (f(n) - f(0))}{n\pi(z - n)}.$$

3. Obtain the cardinal series for functions of the form

$$f(z) = \frac{1}{2\pi} \int_{-\pi}^{\pi} \phi(t) e^{-izt}\, dt,$$

with ϕ integrable on $[-\pi, \pi]$. (*Hint*: Replace ϕ by its Fourier series over $[-\pi, \pi]$ and integrate term-by-term.)

4. Show that the function

$$f(z) = \sum_{n=1}^{\infty} \frac{(-1)^n}{n} \frac{\sin \pi(z - n)}{\pi(z - n)}$$

belongs to the Paley–Wiener space but that $zf(z)$ is not bounded on the real axis (cf. Example 1, Section 4).

5. Establish the corollary to Theorem 19.

3

The Completeness of Sets of Complex Exponentials

In this chapter we shall discuss a few of the fundamental completeness properties of sets of complex exponentials $\{e^{i\lambda_n t}\}$ over a finite interval of the real axis. The most extensive results in this direction were obtained by Paley and Wiener [1934] and Levinson [1940]. At the same time, we will be laying the groundwork for a more penetrating investigation of nonharmonic Fourier series in L^2.

By making an appropriate change of variables, we may consider the system $\{e^{i\lambda_n t}\}$ over an interval $[-A, A]$ symmetric with respect to the origin. The spaces of interest will be $L^p[-A, A]$, for $1 \leqq p < \infty$, and $C[-A, A]$.

As we have seen, for the system $\{e^{i\lambda_n t}\}$ to be incomplete in $C[-A, A]$, it is necessary and sufficient that there exist a nontrivial entire function $f(z)$, zero for every λ_n, and expressible in the form

$$f(z) = \int_{-A}^{A} e^{izt}\, d\omega(t), \tag{1}$$

where $\omega(t)$ is of bounded variation on $[-A, A]$. When $L^p[-A, A]$ is considered, $f(z)$ takes the form

$$f(z) = \int_{-A}^{A} e^{izt}\phi(t)\, dt, \tag{2}$$

with ϕ in $L^q[-A, A]$. Here q denotes the conjugate exponent: $1/p + 1/q = 1$.

The remarks above establish a connection between completeness and uniqueness. If there are enough λ_n's to guarantee that $f(z)$ vanishes identically whenever

$$f(\lambda_n) = 0 \quad (n = 1, 2, 3, \ldots),$$

i.e., if $\{\lambda_n\}$ is a *set of uniqueness* for $f(z)$, then the system of exponentials $\{e^{i\lambda_n t}\}$ will be complete in the corresponding space. Therefore, every theorem on uniqueness generates a theorem on completeness, and conversely.

In $L^2[-\pi, \pi]$, the situation is particularly simple. Here the system of exponentials $\{e^{i\lambda_n t}\}$ is complete whenever $\{\lambda_n\}$ is a set of uniqueness for every function belonging to the Paley–Wiener space.

Clearly, completeness in $C[-A, A]$ implies completeness in $L^p[-A, A]$ for every value of p.

It is often advantageous to permit repetitions among the λ_n. Suppose, for example, that some λ_n, call it λ, is repeated m times. In this case the understanding is that $f(z)$, be it of the form (1) or (2), shall have a zero of multiplicity m at $z = \lambda$. It follows that either $\phi(t)$ or the differential $d\omega(t)$ will be orthogonal to each of the functions

$$e^{i\lambda t}, te^{i\lambda t}, \ldots, t^{m-1}e^{i\lambda t}, \tag{3}$$

and hence all of these functions will be available for the approximation. Conversely, the orthogonality of $\phi(t)$ or $d\omega(t)$ to all functions of this form implies that λ is a zero of multiplicity m of the corresponding entire function $f(z)$. Infinite repetitions among the λ_n are excluded.

For simplicity, we shall suppose in what follows that the λ_n are distinct. The reader will have no difficulty in reformulating all completeness theorems to include functions of the form (3).

We begin with a closer look at the trigonometric system.

1 THE TRIGONOMETRIC SYSTEM

We have seen that the trigonometric system $\{e^{int}\}_{-\infty}^{\infty}$ is complete in every L^p space $(1 \leqq p < \infty)$ over an interval of length 2π. In fact, if $\phi(t)$ is integrable on $[-\pi, \pi]$ and

$$\int_{-\pi}^{\pi} \phi(t)e^{int}\,dt = 0 \quad (n = 0, \pm 1, \pm 2, \ldots),$$

then $\phi(t)$ must vanish almost everywhere.

Once a collection of functions is known to be complete, the next concern should be for economy — are there any superfluous terms?

Definition. *A system of complex exponentials* $\{e^{i\lambda_n t}\}$ *is said to be* **exact** *in* $L^p[-A, A]$ *or in* $C[-A, A]$ *if it is complete, but fails to be complete on the removal of any one term. If the system becomes exact when N terms are* *removed, then we say it has* **excess** *N; if it becomes exact when N terms*

$$e^{i\mu_1 t}, \ldots, e^{i\mu_N t}$$

are adjoined, then we say it has **deficiency** *N.*

As we shall see (Theorem 7), the particular functions $e^{i\mu t}$ added or removed are arbitrary; only their number is important.

Proposition 1. *The trigonometric system is exact in* $L^p[-\pi, \pi]$ *for* $1 \leqq p < \infty$*; in* $C[-\pi, \pi]$ *it has deficiency* 1.

Proof. The first assertion is obvious — the trigonometric system is complete and its elements are mutually orthogonal.

For the second assertion, argue as follows. It is clear that the trigonometric system is incomplete in $C[-\pi, \pi]$ since the only continuous functions that can be uniformly approximated by trigonometric polynomials are those for which $f(\pi) = f(-\pi)$. Let us adjoin the element $e^{i\mu t}$, where μ is not an integer. If $f \in C[-\pi, \pi]$, then for a suitable constant c the values of the function

$$f(t) - ce^{i\mu t}$$

at $t = \pi$ and $t = -\pi$ are the same. The new system is then complete, by virtue of the Weierstrass theorem on trigonometric approximation. Thus in $C[-\pi, \pi]$ the trigonometric system has deficiency 1. ■

It is not difficult to show that if the system $\{e^{i\lambda_n t}\}$ is complete in $L^1[-A, A]$, then its deficiency in $C[-A, A]$ cannot exceed 1 (see Problem 2).

Proposition 2. *The trigonometric system is incomplete in* L^1 *over every interval of length greater than* 2π.

Proof. Indeed, given $\varepsilon > 0$, choose

$$\phi(t) = \begin{cases} -1, & t < -\pi + \varepsilon, \\ 0, & |t| \leqq \pi - \varepsilon, \\ 1, & t > \pi - \varepsilon, \end{cases}$$

and put

$$f(z) = \int_{-\pi-\varepsilon}^{\pi+\varepsilon} \phi(t) e^{izt}\, dt.$$

A straightforward calculation shows that

$$f(z) = \frac{4i}{z} \sin \pi z \, \sin \varepsilon z,$$

and hence $f(n) = 0$ for every integer n. Since ε was arbitrary, the result follows. ■

Consider now the set of functions $\{e^{int}\}_{n=1}^{\infty}$ obtained from the trigonometric system by removing roughly "half" the terms. It seems plausible that the resulting set will be complete in L^p over an interval of length π and over no larger interval. Surprisingly, this is not the case. The set $\{e^{int}\}_{n=1}^{\infty}$ is complete in L^p and in C over every interval of length less than 2π. This is a consequence of the following theorem of Carleman [1922].

Theorem 1. *Let $\{\lambda_n\}$ be a set of positive real numbers and suppose that for some positive number A,*

$$\limsup_{R\to\infty}\left(\frac{1}{\log R}\sum_{\lambda_n<R}\frac{1}{\lambda_n}\right) > \frac{A}{\pi}. \tag{1}$$

Then the set $\{e^{i\lambda_n t}\}$ is complete in $C[-A, A]$.

Proof. We argue by contradiction. If the system were incomplete, then we could find a function $\omega(t)$ of bounded variation on $[-A, A]$ such that the entire function

$$f(z) = \int_{-A}^{A} e^{izt}\, d\omega(t) \tag{2}$$

vanishes at each λ_n but does not vanish identically. It is to be shown that condition (1) forces $f(z)$ to have "too many" zeros, given the restriction that (2) imposes on its growth. The pertinent fact concerning the growth of $f(z)$ is the obvious estimate

$$|f(re^{i\theta})| = O(e^{Ar|\sin\theta|}). \tag{3}$$

Applying Carleman's formula to $f(z)$ in the right half-plane and then using (3), we obtain

$$\begin{aligned}\sum_{\lambda_n<R}\left(\frac{1}{\lambda_n}-\frac{\lambda_n}{R^2}\right) &\leqq \frac{1}{2\pi}\int_1^R\left(\frac{1}{r^2}-\frac{1}{R^2}\right)2Ar\,dr + O(1)\\ &\leqq \frac{A}{\pi}\log R + O(1),\end{aligned}$$

and hence

$$\limsup_{R\to\infty}\left(\frac{1}{\log R}\sum_{\lambda_n<R}\left(\frac{1}{\lambda_n}-\frac{\lambda_n}{R^2}\right)\right) \leqq \frac{A}{\pi}.$$

We complete the proof by showing that

$$\limsup_{R\to\infty}\left(\frac{1}{\log R}\sum_{\lambda_n<R}\frac{1}{\lambda_n}\right) = \limsup_{R\to\infty}\left(\frac{1}{\log R}\sum_{\lambda_n<R}\left(\frac{1}{\lambda_n}-\frac{\lambda_n}{R^2}\right)\right). \tag{4}$$

Evidently, the right-hand side cannot exceed the left. If β is an arbitrary positive number belonging to the interval (0, 1), then for each value of R,

$$\sum_{\lambda_n<R}\left(\frac{1}{\lambda_n}-\frac{\lambda_n}{R^2}\right) \geqq \sum_{\lambda_n<\beta R}\frac{1}{\lambda_n}\left(1-\frac{\lambda_n^2}{R^2}\right) > \sum_{\lambda_n<\beta R}\frac{1}{\lambda_n}(1-\beta^2).$$

It follows that

$$\frac{1}{\log \beta R}\sum_{\lambda_n<\beta R}\frac{1}{\lambda_n} < \frac{1}{1-\beta^2}\frac{1}{\log \beta R}\sum_{\lambda_n<R}\left(\frac{1}{\lambda_n}-\frac{\lambda_n}{R^2}\right)$$

$$= \frac{1}{1-\beta^2}\frac{1}{1+\dfrac{\log\beta}{\log R}}\frac{1}{\log R}\sum_{\lambda_n<R}\left(\frac{1}{\lambda_n}-\frac{\lambda_n}{R^2}\right)$$

and hence that

$$\limsup_{R\to\infty}\left(\frac{1}{\log R}\sum_{\lambda_n<R}\frac{1}{\lambda_n}\right) \leqq \frac{1}{1-\beta^2}\limsup_{R\to\infty}\left(\frac{1}{\log R}\sum_{\lambda_n<R}\left(\frac{1}{\lambda_n}-\frac{\lambda_n}{R^2}\right)\right).$$

Letting $\beta \to 0$, we obtain (4). Conclusion:

$$\limsup_{R\to\infty}\left(\frac{1}{\log R}\sum_{\lambda_n<R}\frac{1}{\lambda_n}\right) \leqq \frac{A}{\pi},$$

which is contrary to assumption. The contradiction proves the theorem. ∎

Corollary. *The system $\{e^{int}\}_{n=1}^{\infty}$ is complete in $C[-A, A]$ whenever $A < \pi$.*

Proof. Since

$$\frac{1}{\log N}\sum_{n=1}^{N}\frac{1}{n} \to 1 \quad \text{as} \quad N \to \infty,$$

the result follows at once from Theorem 1. ∎

The next result also follows readily from Theorem 1. The details are left to the reader.

Theorem 2. *If $\{\lambda_n\}$ is a set of positive real numbers for which*

$$\liminf_{n\to\infty}\frac{n}{\lambda_n} > \frac{A}{\pi},$$

then the system $\{e^{i\lambda_n t}\}$ is complete in $C[-A, A]$.

It is possible to replace Theorem 2 by much sharper results. In Section 6 it will be shown, for example, that if

$$\limsup_{n\to\infty} \frac{n}{\lambda_n} > \frac{A}{\pi},$$

then the conclusion of the theorem still holds.

PROBLEMS

1. Show that the deficiency of a system of complex exponentials $\{e^{i\lambda_n t}\}$ in $L^p[-A, A]$ is a nondecreasing function of p.
2. Prove: if the system $\{e^{i\lambda_n t}\}$ is complete in $L^1[-A, A]$, then its deficiency in $C[-A, A]$ cannot exceed 1. (*Hint:* Suppose that 1 does not belong to $\{e^{i\lambda_n t}\}$. The set of all functions of the form

   $$\int_0^x f(t)\,dt + k,$$

   where $f \in L^1[-A, A]$ and k is a constant, is a dense vector subspace of $C[-A, A]$.)
3. Prove Theorem 2.
4. **(Levinson)** Let $\{\lambda_n\}$ be a set of positive real numbers and let $n(t)$ denote the number of λ_n in the interval $[0, t]$. Then the system $\{e^{i\lambda_n t}\}$ is complete in $C[-\pi D, \pi D]$ if

   $$\limsup_{R\to\infty} \int_1^R \frac{n(t) - Dt}{\sqrt{t}} \left(\frac{1}{t} + \frac{1}{R}\right) dt = \infty.$$

   (*Hint:* If $\{e^{i\lambda_n t}\}$ were not complete, then we could find an entire function $f(z)$, zero for every λ_n, but not identically zero, and expressible in the form

   $$f(z) = \int_{-\pi D}^{\pi D} e^{izt}\,d\omega(t),$$

   where $\omega(t)$ is of bounded variation on $[-\pi D, \pi D]$. Put $g(z) = f(z^2)$. By using Carleman's formula, a contradiction can be obtained in much the same way as in the proof of Theorem 1.)
5. Show that the criterion in Problem 4 is sharp in case $\lim_{n\to\infty} n/\lambda_n$ exists.
6. Show that if

   $$\lambda_n = n - n^{1/2} \quad \text{for} \quad n = 2, 3, 4, \ldots,$$

   then the system $\{e^{i\lambda_n t}\}$ is complete in $C[-\pi, \pi]$.
7. Let $\{\lambda_n\}$ be a sequence of positive real numbers for which

   $$\limsup_{n\to\infty} \frac{n}{\lambda_n} = L.$$

Is it true that

$$\limsup_{R\to\infty}\left(\frac{1}{\log R}\sum_{\lambda_n<R}\frac{1}{\lambda_n}\right)=L?$$

2 EXPONENTIALS CLOSE TO THE TRIGONOMETRIC SYSTEM

In this section we shall discuss the stability of the trigonometric system in L^p under small perturbations of the integers. Our goal is to show that if $1<p<\infty$ and $\{\lambda_n\}$ is a sequence of real or complex numbers for which

$$|\lambda_n-n|\leqq\frac{1}{2p},\qquad n=0,\pm1,\pm2,\ldots,$$

then the system $\{e^{i\lambda_n t}\}$ is complete in $L^p[-\pi,\pi]$ (cf. Kadec's $\frac{1}{4}$-theorem). This result is sharp in the sense that the constant $1/2p$ cannot be replaced by any larger number.

As a measure of the density of a sequence of complex numbers $\{\lambda_n\}$, we denote, as usual, by $n(r)$ the number of points λ_n inside the disk $|z|\leqq r$, and we put

$$N(r)=\int_1^r\frac{n(t)}{t}\,dt.$$

The following theorem is fundamental.

Theorem 3 (Levinson). *The set $\{e^{i\lambda_n t}\}$ is complete in $L^p[-\pi,\pi]$, $1<p<\infty$, whenever*

$$\limsup_{r\to\infty}\left(N(r)-2r+\frac{1}{p}\log r\right)>-\infty.$$

Proof. Suppose to the contrary that the set $\{e^{i\lambda_n t}\}$ fails to be complete in $L^p[-\pi,\pi]$ for some value of p. Then there exists a function ϕ in $L^q[-\pi,\pi]$ such that the entire function

$$f(z)=\int_{-\pi}^{\pi}\phi(t)e^{izt}\,dt$$

vanishes at each λ_n, but does not vanish identically. We may suppose without loss of generality that the norm of ϕ does not exceed 1.

Let ε be a small positive number. Then

$$|f(z)|\leqq\left(\int_{|t|\leqq\pi-\varepsilon}+\int_{\pi-\varepsilon\leqq|t|\leqq\pi}\right)|\phi(t)|e^{-yt}\,dt,$$

and, by an application of Hölder's inequality, we see that

$$|f(z)|\leqq e^{\pi|y|}|y|^{-1/p}(e^{-\varepsilon|y|}+\lambda),\tag{1}$$

where

$$\lambda = \left(\int_{\pi-\varepsilon \leqq |t| \leqq \pi} |\phi(t)|^q \, dt \right)^{1/q}.$$

Clearly, $\lambda \to 0$ as $\varepsilon \to 0$.

If we now apply Jensen's formula to $f(z)$ and then use (1), we obtain

$$N(r) \leqq 2r - \frac{1}{p} \log r + \frac{1}{2\pi} \int_0^{2\pi} \log |e^{-\varepsilon r|\sin\theta|} + \lambda| \, d\theta + O(1) \qquad (2)$$

since the λ_n form a part of all the zeros of $f(z)$. Furthermore, if ε is small enough and r is large enough, then the integrated term on the right-hand side of (2) can be made smaller than any preassigned negative number, and it follows that

$$\limsup_{r\to\infty} \left(N(r) - 2r + \frac{1}{p} \log r \right) = -\infty.$$

The result now follows by contraposition. ∎

Theorem 4 (Levinson). *If $1 < p < \infty$ and $\{\lambda_n\}$ is a sequence of real or complex numbers for which*

$$|\lambda_n| \leqq |n| + \frac{1}{2p}, \qquad n = 0, \pm 1, \pm 2, \ldots, \qquad (3)$$

then the system $\{e^{i\lambda_n t}\}$ is complete in $L^p[-\pi, \pi]$. The constant $1/2p$ is the best possible.

Proof. Condition (3) implies that

$$n(t) \geqq 1 + 2\left[t - \frac{1}{2p} \right] \quad \text{whenever} \quad t > 1$$

(remember that $[x]$ denotes the function "bracket x"). It follows that if $r > 1$, then

$$N(r) = \int_1^r \frac{n(t)}{t} \, dt \geqq 2 \int_1^r \frac{1 + 2[t - 1/2p]}{t} \, dt$$

$$= 2 \int_1^r \frac{t - 1/2p}{t} \, dt - 2 \int_1^r \frac{t - 1/2p - [t - 1/2p] - \frac{1}{2}}{t} \, dt.$$

Since $x - [x] - \frac{1}{2}$ is periodic and has average value zero over each interval of length one, the last integral above remains bounded as $r \to \infty$ (see Problem 1). Accordingly,

$$N(r) \geqq 2r - \frac{1}{p} \log r - c$$

for some constant c and all values of r. Hence Theorem 3 applies, and the system $\{e^{i\lambda_n t}\}$ is complete in $L^p[-\pi, \pi]$.

We complete the proof by showing that if ε is an arbitrary positive number and

$$\lambda_n = \begin{cases} n + \dfrac{1}{2p} + \varepsilon, & n > 0, \\ 0, & n = 0, \\ n - \dfrac{1}{2p} - \varepsilon, & n < 0, \end{cases}$$

then the set $\{e^{i\lambda_n t}\}$ is *not* complete in $L^p[-\pi, \pi]$.

Put $c = 1/2p + \varepsilon$. It is to be shown that the function

$$\phi(t) = \left(\cos \tfrac{1}{2}t\right)^{2c-1} \sin \tfrac{1}{2}t$$

is orthogonal to every element $e^{i\lambda_n t} (n = 0, \pm 1, \pm 2, \ldots)$; clearly, $\phi(t)$ belongs to $L^q[-\pi, \pi]$.

Suppose first that n is positive. By writing $\sin \frac{1}{2}t$ and $\cos \frac{1}{2}t$ in complex form, we obtain

$$\int_{-\pi}^{\pi} \phi(t) e^{i\lambda_n t}\, dt = i2^{-2c} \int_{-\pi}^{\pi} (1 + e^{it})^{2c-1} (1 - e^{it}) e^{int}\, dt.$$

But

$$\begin{aligned} \int_{-\pi}^{\pi} (1 + e^{it})^{2c-1} e^{int}\, dt &= \lim_{r \to 1-} \int_{-\pi}^{\pi} (1 + re^{it})^{2c-1} e^{int}\, dt \\ &= \lim_{r \to 1-} \sum_{k=0}^{\infty} \binom{2c-1}{k} r^k \int_{-\pi}^{\pi} e^{i(n+k)t}\, dt \\ &\qquad \text{(by the binomial theorem)} \\ &= 0 \quad \text{for} \quad n = 1, 2, 3, \ldots. \end{aligned}$$

This shows that $\phi(t)$ is orthogonal to each of the functions $e^{i\lambda_n t}$ (n positive). A similar result holds when n is negative (take conjugates); the case $n = 0$ is obvious. ■

Remarks.

1. Theorem 4 provides simple examples of sets that are complete in $L^p[-\pi, \pi]$ but fail to be complete in $L^r[-\pi, \pi]$ if $r > p$.

2. When $p = 1$, the conclusion of the theorem is no longer valid. If

$$\lambda_n = \begin{cases} n + \frac{1}{2}, & n > 0, \\ 0, & n = 0, \\ n - \frac{1}{2}, & n < 0, \end{cases}$$

then the system $\{e^{i\lambda_n t}\}$ is not complete in $L^1[-\pi, \pi]$. Reason: The function $\sin \frac{1}{2}t$ is orthogonal to every element of the set (verification is left to the reader).

On the other hand, completeness in L^p, $1 < p < \infty$, over a finite interval implies completeness in L^1, and hence the system $\{e^{i\lambda_n t}\}$ is complete in $L^1[-\pi, \pi]$ whenever

$$|\lambda_n - n| \leqq L < \tfrac{1}{2}, \quad n = 0, \pm 1, \pm 2, \ldots.$$

PROBLEMS

1. Show that the integrals

$$\int_1^r \frac{x - [x] - \frac{1}{2}}{x} \, dx$$

remain bounded as $r \to \infty$.

2. Investigate the validity of Theorem 3 when $p = 1$ and when $p = \infty$.

3. **(Levinson)** If $1 < p < \infty$ and if

$$|\lambda_n| \leqq |n| + \frac{1}{2p} + \frac{1}{2}N, \quad n = 0, \pm 1, \pm 2, \ldots,$$

then the set $\{e^{i\lambda_n t}\}$ is either complete in $L^p[-\pi, \pi]$ or else has deficiency at most N.

4. If $\{\varepsilon_n\}_{n=0}^{\infty}$ is a sequence of positive numbers such that $\sum \varepsilon_n/(n+1) < \infty$, then the system $\{e^{i\lambda_n t}\}_{-\infty}^{\infty}$ will be complete in $L^p[-\pi, \pi]$, $1 < p < \infty$, whenever

$$|\lambda_n| \leqq |n| + \frac{1}{2p} + \varepsilon_{|n|} \quad (n = 0, \pm 1, \pm 2, \ldots).$$

3 A COUNTEREXAMPLE

Theorem 4 shows that the set of complex exponentials $\{e^{i\lambda_n t}\}$ is complete in $L^2[-\pi, \pi]$ whenever

$$|\lambda_n - n| \leqq \tfrac{1}{4}, \quad n = 0, \pm 1, \pm 2, \ldots.$$

In the present section we shall prove that this assumption allows the system $\{e^{i\lambda_n t}\}$ to have an *excess*. This will serve to show that Kadec's $\frac{1}{4}$-theorem is "best possible": the system $\{e^{i\lambda_n t}\}$ constitutes a basis for $L^2[-\pi, \pi]$ whenever every λ_n is real and

$$|\lambda_n - n| \leqq L < \tfrac{1}{4}, \quad n = 0, \pm 1, \pm 2, \ldots,$$

but need not constitute a basis when $L = \frac{1}{4}$.

Theorem 5. *The set*

$$\{e^{\pm i(n-\frac{1}{4})t} : n = 1, 2, 3, \ldots\}$$

is exact in $L^2[-\pi, \pi]$ but is not a Riesz basis.

Proof. Put $\lambda_n = n - \frac{1}{4}$ and $\lambda_{-n} = -\lambda_n$ $(n = 1, 2, 3, \ldots)$. The completeness of the set $\{e^{i\lambda_n t}\}$ in $L^2[-\pi, \pi]$ is an almost immediate consequence of Theorem 4: simply replace each λ_n with its translate $\lambda_n + \frac{1}{2}$, and observe that multiplication by $e^{it/2}$ is an isometry on $L^2[-\pi, \pi]$.

Define an entire function $f(z)$ by setting

$$f(z) = a \int_{-\pi}^{\pi} \left(\cos \tfrac{1}{2}t\right)^{-1/2} e^{izt}\, dt,$$

where the constant a is chosen so that $f(0) = 1$. Since $\cos^{-1/2} \frac{1}{2}t$ is integrable on $(-\pi, \pi)$, it follows that $f(z)$ is of exponential type at most π and bounded on the real axis. Assertion:

$$f(\lambda_n) = 0 \quad \text{for every } \lambda_n.$$

The method of proof is the same as that used for a similar purpose in the proof of Theorem 4. If n is positive, then

$$\begin{aligned}
\int_{-\pi}^{\pi} \left(\cos \tfrac{1}{2}t\right)^{-1/2} e^{i\lambda_n t}\, dt &= \sqrt{2} \int_{-\pi}^{\pi} (1 + e^{it})^{-1/2} e^{int}\, dt \\
&= \sqrt{2} \lim_{r \to 1-} \int_{-\pi}^{\pi} (1 + re^{it})^{-1/2} e^{int}\, dt \\
&= \sqrt{2} \lim_{r \to 1-} \sum_{k=0}^{\infty} \binom{-\frac{1}{2}}{k} r^k \int_{-\pi}^{\pi} e^{i(n+k)t}\, dt = 0.
\end{aligned}$$

A similar result holds when n is negative.

Conclusion: For each λ belonging to the set $\{\lambda_n\}$, the function $f(z)/(z - \lambda)$ belongs to the Paley–Wiener space $\boldsymbol{P}$ and vanishes at every other λ_n. This shows that the set $\{e^{i\lambda_n t}\}$ fails to be complete on the removal of any one term and hence it is exact.

It is convenient to complete the proof in three steps.

1. $f(z) = \dfrac{\Gamma^2(\mu)}{\Gamma(\mu + z)\Gamma(\mu - z)}$, where $\mu = 3/4$.

Begin by observing that $f(z)$ can vanish only at the λ_n's. For if $f(z)$ were zero for some other value, say $z = a$, then the function $f(z)/(z - a)$ would belong to $\boldsymbol{P}$ and would vanish at every λ_n. But this is impossible since the completeness of the set $\{e^{i\lambda_n t}\}$ guarantees that $\{\lambda_n\}$ is a set of uniqueness for *every* function belonging to the Paley–Wiener space.

Let

$$g(z) = \prod_{n=1}^{\infty}\left(1 - \frac{z^2}{\lambda_n^2}\right)$$

be the canonical product formed with the zeros $\{\lambda_n\}$. Since the order of $f(z)$ is no greater than one, Hadamard's factorization theorem justifies writing

$$f(z) = e^{Az}g(z).$$

But $f(z)$ and $g(z)$ are both even, and hence A must vanish. Thus $f = g$.

The preceding argument shows that there is a unique entire function of order one that is even, zero only at the λ_n's, and one at the origin. Since the function

$$\frac{\Gamma^2(\mu)}{\Gamma(\mu+z)\Gamma(\mu-z)}$$

satisfies all of these properties, it must be $f(z)$.

2. We require a simple formula for $f'(\lambda_n)$ when n is positive. To begin,

$$f'(z) = \Gamma^2(\mu)\left\{\frac{1}{\Gamma(\mu+z)}\frac{\Gamma'(\mu-z)}{\Gamma^2(\mu-z)} - \frac{1}{\Gamma(\mu-z)}\frac{\Gamma'(\mu+z)}{\Gamma^2(\mu+z)}\right\}$$

and

$$f'(\lambda_n) = \frac{\Gamma^2(\mu)}{\Gamma(\mu+\lambda_n)} \cdot \lim_{z\to\lambda_n}\frac{\Gamma'(\mu-z)}{\Gamma^2(\mu-z)} = \frac{\Gamma^2(\mu)}{\Gamma\left(n+\frac{1}{2}\right)} \cdot \lim_{z\to-n+1}\frac{\Gamma'(z)}{\Gamma^2(z)}.$$

The claim is that

$$\lim_{z\to-n}\frac{\Gamma'(z)}{\Gamma^2(z)} = (-1)^{n+1}n! \quad \text{for} \quad n = 0, 1, 2, \ldots.$$

For $n = 0$, argue as follows. Write

$$\frac{\Gamma'(z)}{\Gamma^2(z)} = \frac{1}{\Gamma(z)}\frac{\Gamma'(z)}{\Gamma(z)} = \frac{\Psi(z)}{\Gamma(z)} = \frac{1}{\Gamma(z)}\left\{-\frac{1}{z} - \gamma + \sum_{n=1}^{\infty}\frac{z}{n(n+z)}\right\}.$$

Here, $\Psi(z) \equiv \Gamma'(z)/\Gamma(z)$ is the *digamma function*, the logarithmic derivative of $\Gamma(z)$. Since $1/z\Gamma(z) \to 1$ as $z \to 0$, the result for $n = 0$ follows. Proceed by induction. Since $\Gamma(z+1) = z\Gamma(z)$ and $\Psi(z+1) = \Psi(z) + 1/z$, we find

$$\begin{aligned}\lim_{z\to-(n+1)}\frac{\Gamma'(z)}{\Gamma^2(z)} &= \lim_{z\to-n}\frac{\Psi(z-1)}{\Gamma(z-1)} = \lim_{z\to-n}\frac{\Psi(z) - 1/(z-1)}{\Gamma(z)/(z-1)} \\ &= \lim_{z\to-n}\left(\frac{\Psi(z)}{\Gamma(z)}(z-1) - \frac{1}{\Gamma(z)}\right) = (-1)^{n+2}(n+1)!.\end{aligned}$$

This proves the claim.

Therefore, when n is positive,

$$f'(\lambda_n) = (-1)^n \Gamma^2\left(\tfrac{3}{4}\right) \frac{\Gamma(n)}{\Gamma\left(n+\frac{1}{2}\right)}.$$

3. The proof of the theorem is now within reach. Suppose to the contrary that the set $\{e^{i\lambda_n t}\}$ *were* a Riesz basis for $L^2[-\pi, \pi]$. Then we could write

$$1 = \sum c_n e^{i\lambda_n t} \quad \textit{(in the mean)}.$$

The sequence of coefficients $\{c_n\}$ is unique and belongs to l^2.

By means of the Fourier isometry, the entire discussion can be transformed into the Paley–Wiener space. The exponentials $e^{i\lambda_n t}$ are transformed into the reproducing functions

$$K_n(z) = \frac{\sin \pi(z - \lambda_n)}{\pi(z - \lambda_n)}$$

and the expansion

$$\frac{\sin \pi z}{\pi z} = \sum c_n K_n(z)$$

is valid in the topology of $\boldsymbol{P}$.

Define a sequence $\{f_n\}$ of functions of $\boldsymbol{P}$ by setting

$$f_n(z) = \frac{f(z)}{f'(\lambda_n)(z - \lambda_n)}.$$

Then $\{f_n\}$ and $\{K_n\}$ are biorthogonal, since

$$(f_n, K_m) = f_n(\lambda_m) = \delta_{mn}.$$

Therefore,

$$c_n = \left(\frac{\sin \pi z}{\pi z}, f_n(z)\right) = f_n(0) = \frac{-1}{\lambda_n f'(\lambda_n)}.$$

Using the asymptotic formula

$$\frac{\Gamma(n)}{\Gamma\left(n+\frac{1}{2}\right)} \sim \frac{1}{\sqrt{n}} \quad \text{as} \quad n \to \infty,$$

we conclude that

$$\sum |c_n|^2 \quad \textit{diverges.}$$

The contradiction completes the proof. ∎

Remark. By refining the argument, we can prove that the set

$$\{e^{\pm i(n-1/4)t} : n = 1, 2, 3, \ldots\}$$

is not even a *basis* for $L^2[-\pi, \pi]$ (see Problem 1).

PROBLEMS

1. Show that the set $\{e^{\pm i(n-1/4)t}: n = 1, 2, 3, \ldots\}$ is not a basis for $L^2[-\pi, \pi]$. (*Hint:* The series

$$\sum c_n e^{i\lambda_n t}$$

that appears in the proof of Theorem 5 *diverges* in $L^2[-\pi, \pi]$.)

2. Show that if

$$g(z) = \prod_{n=1}^{\infty} \left(1 - \frac{z^2}{\lambda_n^2}\right),$$

where $\lambda_n = n - \frac{1}{4}$, then

$$g(x) \sim c \frac{\sin \pi \left(x + \frac{1}{4}\right)}{\sqrt{x}} \quad \text{as} \quad x \to \infty.$$

3. Is the set $\{e^{\pm i(n+1/4)t}: n = 1, 2, 3, \ldots\}$ exact in $L^2[-\pi, \pi]$?

4 SOME INTRINSIC PROPERTIES OF SETS OF COMPLEX EXPONENTIALS

The close interplay between entire functions of the form

$$f(z) = \int_{-A}^{A} e^{izt}\, d\omega(t),$$

where $\omega(t)$ is of bounded variation on the interval $[-A, A]$, and systems of complex exponentials $\{e^{i\lambda_n t}\}$ endows these systems with special intrinsic properties. One of the most useful states that the completeness of such a system is unaffected when one of the terms $e^{i\lambda t}$ is replaced by a different term $e^{i\mu t}$ (different from those already present). The proof of this assertion is particularly transparent in the case of $L^2[-\pi, \pi]$. For if $f(z)$ belongs to the Paley–Wiener space $\boldsymbol{P}$, if

$$f(\lambda_n) = 0 \quad \text{whenever} \quad \lambda_n \neq \lambda$$

and if

$$f(\mu) = 0,$$

then the function

$$g(z) = \frac{z - \lambda}{z - \mu} f(z)$$

also belongs to $\boldsymbol{P}$ and satisfies

$$g(\lambda_n) = 0 \quad \text{for every } \lambda_n.$$

If $\{e^{i\lambda_n t}\}$ is complete, then $\{\lambda_n\}$ is a set of uniqueness for $g(z)$, and so $g(z)$ must vanish identically. It follows that $f(z)$ must also vanish identically, and hence the new system (with $e^{i\lambda t}$ replaced by $e^{i\mu t}$) is also complete.

For L^p spaces in general, the argument above must be modified. The essential ingredient is not the Paley–Wiener theorem, but the fact that if $f(z)$ is the Fourier transform of an L^p function and $f(\mu) = 0$, then $g(z)$ is also the Fourier transform of an L^p function. This is the content of the next theorem.

Theorem 6. *Let*

$$f(z) = \int_{-\pi}^{\pi} \alpha(t)e^{izt}\,dt,$$

where $\alpha(t) \in L^p[-\pi, \pi]$ *and* $1 \leqq p < \infty$. *If* $f(\mu) = 0$ *and*

$$g(z) = \frac{z-\lambda}{z-\mu} f(z),$$

then there exists a function $\beta(t)$ *in* $L^p[-\pi, \pi]$ *such that*

$$g(z) = \int_{-\pi}^{\pi} \beta(t)e^{izt}\,dt. \tag{1}$$

The relation

$$\beta(t) = \alpha(t) + i(\lambda - \mu)e^{-i\mu t} \int_{-\pi}^{t} \alpha(s)e^{i\mu s}\,ds \tag{2}$$

holds almost everywhere on $[-\pi, \pi]$.

Proof. To motivate the proof, let us suppose that $g(z)$ is in fact representable in the form (1), and then try to deduce (2). If (1) holds, then

$$\frac{1}{z-\mu} \int_{-\pi}^{\pi} \alpha(t)e^{izt}\,dt = \frac{1}{z-\lambda} \int_{-\pi}^{\pi} \beta(t)e^{izt}\,dt.$$

The trick in solving for $\beta(t)$ is to transform each of these integrals by first rewriting e^{izt} as

$$e^{izt} = e^{i(z-\mu)t}e^{i\mu t}$$

and then integrating by parts. When this is done, the result is

$$\frac{1}{z-\mu} \int_{-\pi}^{\pi} \alpha(t)e^{izt}\,dt = \int_{-\pi}^{\pi} \alpha_1(t)e^{izt}\,dt,$$

with $\alpha_1(t) = -ie^{-i\mu t} \int_{-\pi}^{t} \alpha(s)e^{i\mu s}\,ds$, and

$$\frac{1}{z-\lambda} \int_{-\pi}^{\pi} \beta(t)e^{izt}\,dt = \int_{-\pi}^{\pi} \beta_1(t)e^{izt}\,dt,$$

with $\beta_1(t) = -ie^{-i\lambda t}\int_{-\pi}^{t}\beta(s)e^{i\lambda s}\,ds$. It follows that

$$\alpha_1(t) = \beta_1(t)$$

almost everywhere on $[-\pi, \pi]$, and so (almost everywhere)

$$e^{i(\lambda-\mu)t}\int_{-\pi}^{t}\alpha(s)e^{i\mu s}\,ds = \int_{-\pi}^{t}\beta(s)e^{i\lambda s}\,ds.$$

To obtain (2), differentiate both sides of this equation with respect to t.

To prove the theorem, simply observe that all of the preceding steps are reversible; clearly, $\beta(t) \in L^p[-\pi, \pi]$. ∎

Remark. A similar result holds when $f(z)$ is of the form

$$f(z) = \int_{-\pi}^{\pi} e^{izt}\,d\alpha(t),$$

and $\alpha(t)$ is of bounded variation on $[-\pi, \pi]$, only now

$$g(z) = \int_{-\pi}^{\pi} e^{izt}\,d\beta(t),$$

with

$$d\beta(t) = d\alpha(t) + i(\lambda - \mu)e^{-i\mu t}\int_{-\pi}^{t} e^{i\mu s}\,d\alpha(s).$$

The following corollary is now immediate.

Theorem 7. *The completeness of the system $\{e^{i\lambda_n t}\}$ in $L^p[-A, A]$, $1 \leqq p < \infty$, or in $C[-A, A]$ is unaffected if some λ_n is replaced by another number.*

For the remainder of this section, we shall restrict attention to *uniform* approximation over a given finite interval of the real axis. The reader will have no difficulty in verifying that all results remain valid for approximations in the L^p metric.

In order that an arbitrary sequence of continuous functions $f_1, f_2, f_3, \ldots$ defined on a closed interval $[a, b]$ be complete in the space $C[a, b]$, it is necessary that *every* element of the space be uniformly approximable by finite combinations of the f_n's. For sequences consisting of complex exponential functions $e^{i\lambda_n t}$, the situation is dramatically simpler: such a system is a fortiori complete provided only that *one other* exponential function $e^{i\lambda t}$ can be so approximated.

Theorem 8. *The system $\{e^{i\lambda_n t}\}$ is complete in $C[a, b]$ if and only if its closed linear span contains one other exponential function $e^{i\lambda t}$.*

Proof. The necessity is trivial. For the sufficiency, suppose that for some number λ (different from every λ_n), the function $e^{i\lambda t}$ can be uniformly approximated on $[a, b]$ by finite linear combinations of the elements $e^{i\lambda_n t}$ $(n = 1, 2, 3, \ldots)$. Then the constant function 1 can be uniformly approximated by finite combinations of the elements $e^{i(\lambda_n - \lambda)t}$. Since $\lambda_n - \lambda$ is never zero, integration shows that all of the powers

$$1, t, t^2, t^3, \ldots$$

can be so approximated. We conclude that for each polynomial $p(t)$ the function

$$e^{i\lambda t} p(t)$$

can be uniformly approximated by finite linear combinations of elements from the set $\{e^{i\lambda_n t}\}$. The proof is over: If f is an arbitrary element of $C[a, b]$, then the function

$$e^{-i\lambda t} f(t)$$

can be uniformly approximated on $[a, b]$ by a polynomial, and the result follows. ■

As a corollary we have the following theorem.

Theorem 9. *Every incomplete set of complex exponentials in* $C[a, b]$ *is minimal.*

(Recall that a set of vectors in a normed vector space is said to be *minimal* if each vector in the set lies outside the closed subspace spanned by the others.)

Definition. *A set of vectors in a normed vector space is said to be* ***linked*** *if each vector in the set lies within the closed subspace spanned by the others.*

Theorem 10. *Every set of complex exponentials in* $C[a, b]$ *is either minimal or linked.*

Proof. If the set $\{e^{i\lambda_n t}\}$ is not complete in $C[a, b]$, then by virtue of Theorem 9 it is minimal, and there is nothing more to be said. If it is complete, then it is either minimal or else some element, say $e^{i\lambda_k t}$, belongs to the closed subspace spanned by the others. In the latter instance, the set $\{e^{i\lambda_n t}\}$ remains complete when the kth term is removed, and hence (by Theorem 7) when *any* term is removed. This shows that each element of the set $\{e^{i\lambda_n t}\}$ belongs to the closed subspace spanned by the others. ■

PROBLEMS

1. **(Paley–Wiener)** The system $\{e^{i\lambda_n t}\}$ is minimal in $L^2[-\pi, \pi]$ if and only if there exists a nontrivial entire function $f(z)$ of exponential type at most π,

zero at every λ_n, and such that

$$\int_{-\infty}^{\infty} \frac{|f(x)|^2}{1+x^2}\, dx < \infty.$$

2. Let $\{\lambda_n\}$ be an increasing sequence of positive real numbers. Show that in $C[0, 1]$ the system $\{t^{\lambda_n}\}$ is either minimal or linked depending on whether the series $\sum 1/\lambda_n$ is convergent or divergent.

3. Prove that the system $\{e^{i\lambda_n t}\}_{-\infty}^{\infty}$ is complete in $L^p[-\pi, \pi]$, $1 < p < \infty$, whenever

$$\sup_n |\operatorname{Re} \lambda_n - n| < \frac{1}{2p} \quad \text{and} \quad \sup_n |\operatorname{Im} \lambda_n| < \infty.$$

5 STABILITY

We have seen that the completeness of the trigonometric system is preserved under sufficiently small perturbations of the integers. It is only natural to ask whether a similar assertion is true for sets of complex exponentials in general. Suppose, for example, that the set $\{e^{i\lambda_n t}\}$ is known to be complete in $L^2[-\pi, \pi]$. Can we find a positive number ε with the property that the set $\{e^{i\mu_n t}\}$ is also complete in $L^2[-\pi, \pi]$ whenever

$$|\lambda_n - \mu_n| \leqq \varepsilon, \quad n = 1, 2, 3, \ldots? \tag{1}$$

The answer is no. Take

$$\lambda_n = \begin{cases} n - \frac{1}{4}, & n > 0, \\ n + \frac{1}{4}, & n < 0. \end{cases}$$

Then the set of exponentials $\{e^{i\lambda_n t}\}$ is complete in $L^2[-\pi, \pi]$ by Theorem 5. Choose $\varepsilon > 0$ and put

$$\mu_n = \begin{cases} \lambda_n + \varepsilon, & n > 0, \\ \lambda_n - \varepsilon, & n < 0. \end{cases}$$

Then $|\lambda_n - \mu_n| = \varepsilon$, and if ε is sufficiently small, then also $\sup_n |\mu_n - n| < \frac{1}{4}$. It follows from Kadec's $\frac{1}{4}$-theorem that the set $\{1, e^{i\mu_n t}\}$ forms a (Riesz) basis for $L^2[-\pi, \pi]$, and hence the removal of a single term (in particular, the element 1) leaves an incomplete set.

One of the ways in which condition (1) may be strengthened is by requiring that

$$|\lambda_n - \mu_n| \leqq \varepsilon_n, \quad n = 1, 2, 3, \ldots,$$

and then imposing suitable restrictions on the ε_n. It is always sufficient to take $\{\varepsilon_n\}$ in l^1.

Theorem 11. *Let $\{\lambda_n\}$ and $\{\mu_n\}$ be two sequences of real numbers for which*

$$\sum_{n=1}^{\infty} |\lambda_n - \mu_n| < \infty$$

and let $1 \leqq p < \infty$. If $\{e^{i\lambda_n t}\}$ is complete in $L^p[-A, A]$, then $\{e^{i\mu_n t}\}$ is also complete in $L^p[-A, A]$.

Proof. Suppose to the contrary that $\{e^{i\mu_n t}\}$ is not complete in $L^p[-A, A]$. Then there exists an entire function $f(z)$, equal to zero for every μ_n, but not identically zero, and expressible in the form

$$f(z) = \int_{-A}^{A} \phi(t) e^{izt}\, dt,$$

with $\phi(t)$ in $L^q[-A, A]$.

Motivated by Theorem 6, we define a sequence of functions $\phi_0, \phi_1, \phi_2, \ldots$ recursively by first setting $\phi_0 = \phi$ and then taking

$$\phi_n(t) = \phi_{n-1}(t) + i(\lambda_n - \mu_n) e^{-i\mu_n t} \int_{-A}^{t} \phi_{n-1}(s) e^{i\mu_n s}\, ds \quad (n = 1, 2, 3, \ldots). \tag{2}$$

Put

$$f_n(z) = \int_{-A}^{A} \phi_n(t) e^{izt}\, dt.$$

Then

$$f_n(z) = \frac{z - \lambda_n}{z - \mu_n} f_{n-1}(z) \quad (n = 1, 2, 3, \ldots) \tag{3}$$

by virtue of Theorem 6, and consequently $f_n(z) = 0$ whenever z is one of the points

$$\lambda_1, \ldots, \lambda_n, \mu_{n+1}, \mu_{n+2}, \ldots.$$

Assertion: $\{\phi_n\}$ is a Cauchy sequence in $L^q[-A, A]$. To prove this, argue as follows. Since each μ_n is real, (2) implies that

$$\|\phi_n - \phi_{n-1}\| \leqq 2A|\lambda_n - \mu_n| \cdot \|\phi_{n-1}\|$$

or

$$\|\phi_n - \phi_{n-1}\| \leqq \varepsilon_n \|\phi_{n-1}\|, \tag{4}$$

where $\sum \varepsilon_n < \infty$. It follows that

$$(1 - \varepsilon_n)\|\phi_{n-1}\| \leqq \|\phi_n\| \leqq (1 + \varepsilon_n)\|\phi_{n-1}\|. \tag{5}$$

Notice that the right-hand side of (5) shows that the sequence of norms $\|\phi_n\|$, $n = 1, 2, 3, \ldots$, is bounded: indeed,

$$\|\phi_n\| \leqq (1 + \varepsilon_1) \cdots (1 + \varepsilon_n)\|\phi\| \leqq C\|\phi\|,$$

where $C = \prod_{n=1}^{\infty}(1+\varepsilon_n) < \infty$. Therefore, by virtue of (4), we have

$$\begin{aligned}\|\phi_{n+p} - \phi_n\| &\leqq \|\phi_{n+1} - \phi_n\| + \cdots + \|\phi_{n+p} - \phi_{n+p-1}\| \\ &\leqq \varepsilon_{n+1}\|\phi_n\| + \cdots + \varepsilon_{n+p}\|\phi_{n+p-1}\| \\ &\leqq C\|\phi\|(\varepsilon_{n+1} + \cdots + \varepsilon_{n+p}),\end{aligned}$$

which tends to zero as $n, p \to \infty$ since $\sum \varepsilon_n < \infty$. This proves the assertion.

Accordingly, there is a function ψ in $L^q[-A, A]$ such that $\phi_n \to \psi$ in $L^q[-A, A]$. Since no ϕ_n is trivial (because of (3)), it follows that ψ is not trivial (because of (5)). Put

$$g(z) = \int_{-A}^{A} \psi(t)e^{izt}\,dt.$$

Since

$$f_n(z) \to g(z) \quad \text{as} \quad n \to \infty,$$

we conclude that

$$g(\lambda_n) = 0 \quad \text{for every } \lambda_n,$$

and hence that $\{e^{i\lambda_n t}\}$ is incomplete in $L^q[-A, A]$. The contradiction proves the theorem. ■

Theorem 11 provides a stability criterion for systems of complex exponentials $\{e^{i\lambda_n t}\}$ whenever the λ_n undergo a suitable "horizontal" displacement. Under a "vertical" displacement, much more can be said (at least in L^2).

Theorem 12. *Let $\{\lambda_n\}$ and $\{\mu_n\}$ be two sequences of points lying in a fixed horizontal strip, and suppose that*

$$\operatorname{Re}\lambda_n = \operatorname{Re}\mu_n.$$

If $\{e^{i\lambda_n t}\}$ is complete in $L^2[-\pi, \pi]$, then $\{e^{i\mu_n t}\}$ is also complete in $L^2[-\pi, \pi]$.

Proof. We may assume that neither set $\{\lambda_n\}$ nor $\{\mu_n\}$ contains the point $z = 0$ (why?). Arguing by contradiction, we suppose that $\{e^{i\mu_n t}\}$ is not complete in $L^2[-\pi, \pi]$, so that there exists an entire function $f(z)$, zero for every μ_n, but not identically zero, and expressible in the form

$$f(z) = \int_{-\pi}^{\pi} \phi(t)e^{izt}\,dt,$$

where $\phi \in L^2[-\pi, \pi]$. In addition, we may assume that $f(0) = 1$ (divide $f(z)$ by a suitable power of z and then invoke Theorem 6).

We define a sequence of functions $f_0, f_1, f_2, \ldots$ as follows:

$$f_0(z) = f(z) \quad \text{and} \quad f_n(z) = f(z)\prod_{k=1}^{n}\frac{1 - z/\lambda_k}{1 - z/\mu_k} \qquad (n = 1, 2, 3, \ldots).$$

If $n \geqq 1$, then each $f_n(z)$ belongs to the Paley–Wiener space $\boldsymbol{P}$, assumes the value 1 at $z = 0$, and satisfies

$$f_n(\lambda_k) = 0 \quad (k = 1, \ldots, n).$$

We are going to show that the norms

$$\|f_n\| = \left(\int_{-\infty}^{\infty} |f_n(x)|^2\, dx\right)^{1/2}$$

have a uniform upper bound for all values of n. For this purpose, write

$$f_n(z) = \int_{-\pi}^{\pi} \phi_n(t) e^{izt}\, dt,$$

with ϕ_n in $L^2[-\pi, \pi]$. Since

$$f_n(z) = \frac{\mu_n}{\lambda_n} \frac{z - \lambda_n}{z - \mu_n} f_{n-1}(z)$$

(by definition), Theorem 6 implies that

$$\phi_n(t) = \frac{\mu_n}{\lambda_n}\left(\phi_{n-1}(t) + i(\lambda_n - \mu_n) e^{-i\mu_n t} \int_{-\pi}^{t} \phi_{n-1}(s) e^{i\mu_n s}\, ds\right).$$

Since $\{\operatorname{Im} \lambda_n\}$ and $\{\operatorname{Im} \mu_n\}$ are both bounded, obvious estimates show that

$$\|\phi_n\| \leqq A \frac{|\mu_n|}{|\lambda_n|} \|\phi_{n-1}\|$$

for some constant A and all values of n, and hence

$$\|\phi_n\| \leqq A\|\phi\| \cdot \left|\frac{\mu_1 \cdots \mu_n}{\lambda_1 \cdots \lambda_n}\right|.$$

But the products

$$\left|\frac{\mu_1 \cdots \mu_n}{\lambda_1 \cdots \lambda_n}\right|$$

have a uniform upper bound for all values of n. Indeed,

$$\left|\frac{\mu_k}{\lambda_k}\right|^2 = 1 + \frac{(\operatorname{Im} \mu_k)^2 - (\operatorname{Im} \lambda_k)^2}{|\lambda_k|^2} \leqq 1 + \frac{B}{|\lambda_k|^2}$$

for some constant B and $k = 1, 2, 3, \ldots$, and so

$$\left|\frac{\mu_1 \cdots \mu_n}{\lambda_1 \cdots \lambda_n}\right|^2 \leqq \prod_{k=1}^{\infty} \left(1 + \frac{B}{|\lambda_k|^2}\right). \tag{6}$$

Since $f(z)$ is an entire function of order not exceeding 1, the exponent of convergence of its zeros is also no larger than 1, and so, in particular,

$$\sum_{k=1}^{\infty} \frac{1}{|\mu_k|^2} < \infty.$$

This shows that the series $\sum_{k=1}^{\infty} 1/|\lambda_k|^2$ is also convergent, and hence that the infinite product in (6) is finite. It follows that

$$\sup_n \|\phi_n\| < \infty,$$

and hence also that

$$\sup_n \|f_n\| < \infty,$$

since the Fourier transform is an isometry.

Since $\boldsymbol{P}$ is a Hilbert space, a subsequence of $\{f_n\}$ will converge weakly in $\boldsymbol{P}$ to an element g of $\boldsymbol{P}$. This means that

$$(f_{n_k}, h) \to (g, h) \quad \text{as} \quad k \to \infty$$

for a suitable sequence of integers $\{n_k\}$ and all h in $\boldsymbol{P}$. Choosing h to be the *reproducing function* at z, we find

$$f_{n_k}(z) \to g(z) \quad \text{as} \quad k \to \infty$$

for every z. In particular,

$$g(0) = 1 \quad \text{and} \quad g(\lambda_n) = 0 \quad \text{for} \quad n = 1, 2, 3, \ldots.$$

The proof is over. We have established the existence of a nontrivial function in $\boldsymbol{P}$ that vanishes at every λ_n. Conclusion: The system $\{e^{i\lambda_n t}\}$ is *not* complete in $L^2[-\pi, \pi]$. The contradiction proves the theorem. ■

We have thereby obtained the following strengthening of Theorem 4.

Corollary. *If $\{\lambda_n\}$ is a sequence of complex numbers for which*

$$|\operatorname{Re} \lambda_n| \leqq |n| + \tfrac{1}{4} \quad \textit{and} \quad |\operatorname{Im} \lambda_n| \leqq M \quad (n = 0, \pm 1, \pm 2, \ldots),$$

then the system $\{e^{i\lambda_n t}\}$ is complete in $L^2[-\pi, \pi]$.

PROBLEMS

1. (Redheffer) If

$$\sum \frac{|\lambda_n - \mu_n|}{1 + |\operatorname{Im} \lambda_n| + |\operatorname{Im} \mu_n|} < \infty,$$

then the systems $\{e^{i\lambda_n t}\}$ and $\{e^{i\mu_n t}\}$ have the same excess or deficiency in $L^p[-A, A]$ for $1 \leqq p < \infty$. (*Hint*: If ϕ_n is defined as in the proof of Theorem 11, then

$$\|\phi_n - \phi_{n-1}\| \leqq \|\phi_{n-1}\| \cdot |\lambda_n - \mu_n| \cdot \min\left(2A, \frac{1}{|\operatorname{Im}\lambda_n|}\right).)$$

2. Does the conclusion of Problem 1 still hold when $L^p[-A, A]$ is replaced by $C[-A, A]$?

3. Prove that the condition

$$|\lambda_n - n| < \tfrac{1}{4}, \quad n = 0, \pm 1, \pm 2, \ldots,$$

is not strong enough to guarantee that the system of exponentials $\{e^{i\lambda_n t}\}$ is a basis for $L^2[-\pi, \pi]$ (cf. Kadec's $\frac{1}{4}$-theorem).

6 DENSITY AND THE COMPLETENESS RADIUS

For systems of complex exponentials $\{e^{i\lambda_n t}\}$, the dichotomy between completeness and incompleteness, relative to a given space, is exceedingly complex. Thus, while the trigonometric system is incomplete in $C[-\pi, \pi]$, it becomes complete on the addition of a single arbitrary term $e^{i\lambda t}$.

An important measure of completeness, far less delicate than the algebraic notions of excess or deficiency, is that of the *completeness radius.*

Definition. *Let $\Lambda = \{\lambda_n\}$ be a sequence of real or complex numbers. The* ***completeness radius*** *of Λ is defined to be the number*

$$R(\Lambda) = \sup\{A\colon \{e^{i\lambda_n t}\} \text{ is complete in } C[-A, A]\}.$$

By convention, $R(\Lambda) = 0$ if the set $\{e^{i\lambda_n t}\}$ fails to be complete over any interval $[-A, A]$ and $R(\Lambda) = \infty$ if it is complete over every finite interval $[-A, A]$.

It is not difficult to show that the value of $R(\Lambda)$ remains the same if a finite number of points are removed from or adjoined to Λ. Moreover, the metric in the definition may be replaced by any L^p metric, or by a variety of other topologies, and the value of $R(\Lambda)$ does not change (see Problems 2 and 5).

For a given sequence $\{\lambda_n\}$, the size of the corresponding completeness radius is closely related to the behavior of the ratios $n(r)/r$ as $r \to \infty$. (Here, as usual, $n(r)$ denotes the counting function of $\{\lambda_n\}$, the number of λ_n for which $|\lambda_n| \leqq r$.) We shall say that $\{\lambda_n\}$ has **density** D $(0 \leqq D \leqq \infty)$ if

$$\lim_{r\to\infty} \frac{n(r)}{r} = D.$$

If the λ_n are arranged in nondecreasing order of magnitude, then it follows readily from the definition that $\{\lambda_n\}$ has density D if and only if

$$\lim_{n\to\infty} \frac{n}{|\lambda_n|} = D.$$

A sequence that possesses a density will be called a **measurable sequence**.

The following theorem is simply a reformulation of Theorem 2 under the condition that $\{\lambda_n\}$ is measurable.

Theorem 13. *If $\{\lambda_n\}$ is a sequence of positive numbers with density D and completeness radius R, then $R \geqq \pi D$.*

Suppose now that $\{\lambda_n\}$ is a (not necessarily measurable) sequence of positive numbers, arranged so that $0 < \lambda_1 < \lambda_2 < \cdots$. It was shown by Pólya [1929] that of all the measurable sequences that contain $\{\lambda_n\}$ as a subsequence, there is one whose density is a minimum. This smallest density is called the **Pólya maximum density** of the sequence $\{\lambda_n\}$; it is equal to

$$\lim_{\zeta\to 1-} \limsup_{r\to\infty} \frac{n(r) - n(\zeta r)}{r - \zeta r}.$$

It is clear that the Pólya maximum density is at least as large as the "upper density" $\limsup_{r\to\infty} n(r)/r$.

We state without proof (see Problem 8) the following impressive generalization of Theorem 13.

Theorem 14 (Levinson). *If $\{\lambda_n\}$ is a sequence of positive numbers with Pólya maximum density D and completeness radius R, then $R \geqq \pi D$.*

Example. Consider the sequence $\{\lambda_n\}$ consisting of all integers m for which

$$2^k < m \leqq 2^k + 2^{k-1} \quad (k = 1, 2, 3, \ldots).$$

Then $\{\lambda_n\}$ consists of the numbers

$$\underbrace{3,}\quad \underbrace{5, 6,}\quad \underbrace{9, 10, 11, 12,}\quad \ldots.$$

It is easy to see that $\{\lambda_n\}$ is a nonmeasurable sequence with Pólya maximum density one, and so has completeness radius $R = \pi$. Indeed, the sequence has density one in every interval $(2^k, 2^k + 2^{k-1})$, $k = 1, 2, 3, \ldots$. It is precisely the density in these subintervals that determines the completeness properties of the system $\{e^{i\lambda_n t}\}$.

Notice also that since

$$\liminf_{n\to\infty} \frac{n}{\lambda_n} = \frac{1}{2},$$

Theorem 2 implies only that $R \geqq \pi/2$. The strength of Theorem 14 is thereby forcefully presented.

In the early studies of nonharmonic Fourier series, it was well known that for a large class of sequences having density D the corresponding completeness radius is always equal to πD. Included in this category are those real sequences $\{\lambda_n\}_{-\infty}^{\infty}$ that are *symmetric* ($\lambda_{-n} = -\lambda_n$, $n = 0, 1, 2, \dots$) and "close" to the integers in the sense that

$$|\lambda_n - n| \leqq L, \quad n = 0, \pm 1, \pm 2, \dots$$

(Paley and Wiener [1934, p. 94]). The conjecture by Schwartz [1943, p. 130] that *every* real symmetric sequence having density D must also have $R = \pi D$ therefore seemed eminently reasonable. The conjecture proved false. It was finally and dramatically refuted by Kahane [1958], who constructed a real symmetric sequence for which $D = 0$ and $R = \infty$. Equally surprising was the result of Koosis [1960] showing that there are subsets of the positive integers with $D = 0$ and $R = \pi$.

Nevertheless, for a large class of sequences having density zero, one can conclude that the corresponding completeness radius must also be zero.

Theorem 15. *If $\Lambda = \{\lambda_n\}$ is a sequence of real numbers such that $\sum 1/|\lambda_n| < \infty$, then $R(\Lambda) = 0$.*

The proof will require the following lemma.

Lemma. *Let $\{\lambda_n\}$ be a sequence of real numbers such that*

$$C = \sum_{n=1}^{\infty} \frac{1}{|\lambda_n|} < \infty.$$

If

$$f(z) = \prod_{n=1}^{\infty} \frac{\sin(\pi z/\lambda_n)}{\pi z/\lambda_n}, \tag{1}$$

then $f(z)$ is a nontrivial entire function of exponential type at most πC, bounded by 1 on the real axis, and zero at every λ_n.

Proof. Observe first that since the function $1 - (\sin \pi z)/\pi z$ has a zero at the origin, there exists a constant B such that

$$\left| 1 - \frac{\sin \pi z}{\pi z} \right| \leqq B|z| \quad \text{whenever } |z| \leqq 1.$$

Accordingly,

$$\left| 1 - \frac{\sin(\pi z/\lambda_n)}{\pi z/\lambda_n} \right| \leqq B \left| \frac{z}{\lambda_n} \right| \quad \text{for} \quad |z| \leqq |\lambda_n| \quad (n = 1, 2, 3, \dots).$$

Since $|\lambda_n| \to \infty$, it follows that the series

$$\sum_{n=1}^{\infty} \left| 1 - \frac{\sin(\pi z/\lambda_n)}{\pi z/\lambda_n} \right|$$

converges uniformly on each bounded subset of the plane (apply the Weierstrass M-Test). This shows that the infinite product in (1) defines an entire function $f(z)$ that vanishes at every λ_n but does not vanish identically.

The simple estimate

$$\left| \frac{\sin \pi z}{\pi z} \right| \leqq e^{\pi|z|}$$

implies that

$$|f(z)| \leqq e^{\pi C|z|},$$

so that $f(z)$ is of exponential type at most πC. When x is real, $|(\sin \pi x)/\pi x| \leqq 1$, and hence $|f(x)| \leqq 1$. ∎

It is worth pointing out that we could not have taken $f(z)$ to be the canonical product

$$P(z) = \prod_{n=1}^{\infty} \left(1 - \frac{z}{\lambda_n} \right).$$

Indeed, $P(z)$ is of exponential type zero (by Theorem 2.7), and so cannot be bounded along the real axis (apply Bernstein's inequality).

The proof of Theorem 15 is now readily established.

Proof of Theorem 15. Let A be an arbitrary positive number. We are going to show that there exists an entire function $f(z)$, zero for every λ_n, but not identically zero, and expressible in the form

$$f(z) = \int_{-A}^{A} \phi(t) e^{izt}\, dt,$$

with $\phi(t)$ in $C[-A, A]$. This will show that the corresponding system of exponentials $\{e^{i\lambda_n t}\}$ is incomplete in $L^p[-A, A]$ for $1 \leqq p < \infty$ and hence that $R(\Lambda) = 0$.

Choose N so large that

$$\sum_{n>N} \frac{1}{|\lambda_n|} < \frac{A}{\pi},$$

and define

$$g(z) = \prod_{n>N} \frac{\sin(\pi z/\lambda_n)}{\pi z/\lambda_n}.$$

It follows from the lemma that $g(z)$ is a nontrivial entire function of exponential type at most A, bounded on the real axis, and zero for $z = \lambda_n$ $(n = N+1, N+2, N+3, \ldots)$. Let $\{z_1, \ldots, z_M\}$ be a set of $M = N+2$ zeros of $g(z)$, none of which belongs to the sequence $\{\lambda_n\}$, and form the rational function

$$r(z) = \frac{\prod_{n=1}^{N}(z-\lambda_n)}{\prod_{n=1}^{M}(z-z_n)}.$$

Evidently,

$$r(z) = O\left(\frac{1}{|z|^2}\right) \quad \text{as} \quad |z| \to \infty.$$

Put

$$f(z) = r(z)g(z).$$

Then $f(z)$ is an entire function of exponential type at most A, zero at every λ_n, and not identically zero; in addition, $f(x) \in L^1 \cap L^2$ along the real axis. By the Paley–Wiener theorem, we can write

$$f(z) = \int_{-A}^{A} \phi(t) e^{izt}\, dt,$$

with $\phi \in L^2[-A, A]$. But the Fourier transform of an L^1 function is continuous, and hence $\phi \in C[-A, A]$. ■

The far-reaching study of the completeness radius culminated after more than thirty years in the celebrated work of Beurling and Malliavin [1967]. In a complete solution to the problem, these authors showed that for an arbitrary sequence Λ of real or complex numbers, the corresponding completeness radius $R(\Lambda)$ is proportional to a suitably defined "density" of Λ. The proof, requiring new and deep properties of harmonic and entire functions, is well beyond the scope of the present work.

Research on the broader question of completeness over an interval of length $2R(\Lambda)$ flourishes.

PROBLEMS

1. Let $n(r)$ denote the counting function of a real increasing sequence $\lambda_1, \lambda_2, \lambda_3, \ldots$. Show that

$$\limsup_{r\to\infty} \frac{n(r)}{r} = \limsup_{n\to\infty} \frac{n}{\lambda_n}.$$

Show also that a similar equality holds when lim sup is replaced by lim inf.

2. Show that the value of the completeness radius does not change if in the definition we replace $C[-A, A]$ by any one of the spaces $L^p[-A, A]$ with $1 \leqq p < \infty$.

3. Show that the completeness radius of the sequence $\{\lambda_n\}$ is equal to the greatest lower bound of positive numbers τ for which there exists an entire function of exponential type τ, bounded on the real axis, and zero for every λ_n.

4. Prove that if the sequence $\{\lambda_n\}$ has completeness radius R, then the system of exponentials $\{e^{i\lambda_n t}\}$ has infinite excess in $C[a, b]$ whenever $b - a < 2R$ and infinite deficiency whenever $b - a > 2R$.

5. (a) Let $\{\lambda_1, \lambda_2, \lambda_3, \ldots\}$ and $\{\mu_1, \mu_2, \mu_3, \ldots\}$ have completeness radii R_1 and R_2, respectively. Show that $\{\lambda_1, \mu_1, \lambda_2, \mu_2, \ldots\}$ has completeness radius no larger than $R_1 + R_2$.

(b) Show that the value of R is unaffected by the addition or removal of any number of (real) terms λ_n for which $\sum 1/|\lambda_n| < \infty$.

6. Give an example of a sequence for which the upper density is arbitrarily small and yet the Pólya maximum density is *infinite*.

7. For the example following Theorem 14, compute the value of the upper density of $\{\lambda_n\}$.

8. Prove Theorem 14 by assuming as known the following important but deep fact: the zeros of an entire function of exponential type $\tau > 0$ that is bounded on the real axis have an equal density τ/π in both the right-hand and left-hand planes (Levinson [1940, Theorem VIII]).

9. Show that if $\Lambda = \{\lambda_n\}$ is a sequence of complex numbers for which

$$\sum_{n=1}^{\infty} \frac{|\operatorname{Im} \lambda_n|}{|\lambda_n|^2} = \infty,$$

then $R(\Lambda) = \infty$.

10. Determine the completeness radius of the sequence $\{2, 3, 5, 7, \ldots\}$ consisting of all prime numbers.

4

INTERPOLATION AND BASES IN HILBERT SPACE

Interpolation and approximation often appear as two sides of the same coin; results about the one frequently imply results about the other.

In the present chapter we shall exploit this duality to gain added information about Riesz bases in Hilbert space. This will then serve as the basis for a more penetrating analysis of nonharmonic Fourier series in $L^2[-\pi, \pi]$.

Interpolation problems in an abstract Hilbert space have as their prototype the following classical *trigonometric moment problem* in $L^2[-\pi, \pi]$:

$$\frac{1}{2\pi}\int_{-\pi}^{\pi} \phi(t)e^{-int}\,dt = c_n \quad (n = 0, \pm 1, \pm 2, \ldots).$$

Here the c_n's are given complex numbers and a solution ϕ in $L^2[-\pi, \pi]$ is sought. As is well known, a solution will exist when and only when the scalar sequence $\{c_n\}$ belongs to l^2. The *unique* solution — unique because the trigonometric system is complete — is then given by

$$\phi(t) = \sum_{n=-\infty}^{\infty} c_n e^{int}.$$

In its logical extension to an abstract Hilbert space H, the aforementioned trigonometric moment problem takes the form

$$(f, f_n) = c_n \quad (n = 1, 2, 3, \ldots).$$

Here $f_1, f_2, f_3, \ldots$ are vectors belonging to H, $c_1, c_2, c_3, \ldots$ are scalars, and $f \in H$ is to be found. Two special classes of sequences $\{f_n\}$ will play a prominent role: (1) those for which

$$\sum_{n=1}^{\infty} |(f, f_n)|^2 < \infty$$

whenever $f \in H$ (*Bessel sequences*), and (2) those for which the interpolation problem

$$(f, f_n) = c_n \quad (n = 1, 2, 3, \ldots)$$

is solvable whenever $\{c_n\} \in l^2$ (*Riesz–Fischer sequences*).

The first detailed investigation of sequences satisfying a "Bessel" property or a "Riesz–Fischer" property was made by Bari [1951]. Shared by every orthonormal basis, these two properties are—as we shall subsequently show—the critical ingredients of every Riesz basis.

1 MOMENT SEQUENCES IN HILBERT SPACE

Let H be a Hilbert space and $\{f_1, f_2, f_3, \ldots\}$ a fixed but arbitrary sequence of vectors from H. By the nth **moment** of an element $f \in H$ we mean the inner product (f, f_n). The sequence

$$\{(f, f_n)\}_{n=1}^{\infty}$$

is called the **moment sequence** of f, and the collection of all moment sequences is called the **moment space** of $\{f_n\}$. The moment space will be denoted by M.

It is clear that M is a vector space under pointwise addition and scalar multiplication.

We shall be concerned with the following questions:

(i) When does a given sequence of scalars belong to M, i.e., under what conditions on $\{c_n\}$ does the moment problem

$$(f, f_n) = c_n \quad (n = 1, 2, 3, \ldots) \tag{1}$$

admit at least one solution?

(ii) If there is a solution, is it unique?

(iii) How can the solutions, if any, be recaptured from $\{c_n\}$?

Of these questions, uniqueness alone is trivial: in order that the system of equations (1) shall admit *at most* one solution for every choice of the scalars $c_1, c_2, c_3, \ldots$, it is necessary and sufficient that $\{f_n\}$ be *complete*.

Proposition 1. *Let $\{f_1, f_2, f_3, \ldots\}$ be a sequence of vectors belonging to a Hilbert space H. If for some sequence of scalars $\{c_1, c_2, c_3, \ldots\}$ the system (1) has a solution $f \in H$, then it has a unique solution of minimal norm.*

Proof. Let Y be the closed subspace of H spanned by the elements $f_1, f_2, f_3, \ldots$. Then Y contains *at most one* solution to the moment problem. For if f and g were solutions, each belonging to Y, then $f - g$ would belong to both Y and $Y^{\perp}$. Since $Y \cap Y^{\perp} = \{0\}$, $f = g$. To see that Y contains *at least one* solution to the moment problem, let f be an arbitrary solution in H and write

$$f = \alpha + \beta,$$

where $\alpha \in Y$ and $\beta \in Y^{\perp}$. Clearly $(\alpha, f_n) = c_n$ for every n. Since

$$\|\alpha\| < \|f\|,$$

unless $f = \alpha$, it follows that α is the unique solution of minimal norm. ■

Example 1 (Finite Interpolation). Let $\{f_1, \ldots, f_n\}$ be a linearly independent set of vectors from H. Then the finite moment problem

$$(f, f_i) = c_i \quad (i = 1, \ldots, n)$$

has at least one solution $f \in H$ for every choice of the scalars $c_1, \ldots, c_n$.

This follows easily from the fact that the Gram determinant

$$G = \det((f_i, f_j))_{i,j=1}^{n}$$

of an independent set is never zero. Indeed, the (unique) solution of minimal norm is then given explicitly by the determinantal formula

$$f = -\frac{1}{G} \det \begin{pmatrix} 0 & f_1 & \ldots & f_n \\ c_1 & (f_1, f_1) & \ldots & (f_n, f_1) \\ \vdots & \vdots & & \vdots \\ c_n & (f_1, f_n) & \ldots & (f_n, f_n) \end{pmatrix}.$$

Example 2. The moment space of a Riesz basis is l^2. For if $\{f_n\}$ is a Riesz basis for a separable Hilbert space H, then it possesses a unique biorthogonal sequence $\{g_n\}$ which, by duality, is also a Riesz basis. If $\{c_n\} \in l^2$, then the series

$$\sum_{n=1}^{\infty} c_n g_n$$

converges to an element $f \in H$ which clearly satisfies (1). Accordingly, $M \supset l^2$. On the other hand, the expansion

$$f = \sum_{n=1}^{\infty} (f, f_n) g_n$$

is valid for every element of H, so that $\{(f, f_n)\} \in l^2$ and hence $M \subset l^2$. Thus $M = l^2$.

Example 3. Let $\{\lambda_n\}$ be a symmetric sequence ($\lambda_{-n} = -\lambda_n$) of real numbers, and consider the following "generalized" trigonometric moment problem in $L^2[-\pi, \pi]$:

$$\frac{1}{2\pi}\int_{-\pi}^{\pi} \phi(t)e^{-i\lambda_n t}\,dt = c_n \quad (n = 0, \pm 1, \pm 2, \ldots). \tag{2}$$

If $\{e^{i\lambda_n t}\}$ is in some sense "close" to the trigonometric system, then (2) can be solved explicitly for ϕ.

We shall suppose that the system $\{e^{i\lambda_n t}\}$ forms a Riesz basis for $L^2[-\pi, \pi]$. Then its moment space is l^2, and (2) admits a solution whenever $\{c_n\}$ is square-summable, and for these sequences only. If $\{g_n\}$ is the unique sequence in $L^2[-\pi, \pi]$ biorthogonal to $\{e^{i\lambda_n t}\}$, then the unique solution to (2) is given by the norm-convergent series

$$\phi(t) = \sum_{n=-\infty}^{\infty} c_n g_n(t).$$

We shall determine the g_n's explicitly. For this purpose the following theorem will be needed.

Theorem 1. *Let $\{\lambda_n\}$ be a symmetric sequence of real or complex numbers. If the system of exponentials $\{e^{i\lambda_n t}\}$ is exact in $L^2[-\pi, \pi]$, then the product*

$$\prod_{n=1}^{\infty}\left(1 - \frac{z^2}{\lambda_n^2}\right)$$

converges to an entire function which belongs to the Paley–Wiener space.

Proof. If the system $\{e^{i\lambda_n t}\}$ is exact, then there exists an entire function $f(z)$ which belongs to the Paley–Wiener space $\boldsymbol{P}$ such that $f(\lambda_0) = 1$ and $f(\lambda_n) = 0$ for $n \neq 0$. By Theorem 2.18 we can write

$$f(z) = \frac{1}{2\pi}\int_{-\pi}^{\pi} \varphi(t)e^{izt}\,dt,$$

with $\varphi \in L^2[-\pi, \pi]$. Since $\{\lambda_n\}$ is symmetric, $\varphi(-t)$ has the same orthogonality properties as $\varphi(t)$. But $\varphi(t)$ is unique (because $\{e^{i\lambda_n t}\}$ is complete) and so must be even. Hence $f(z)$ is even.

Now $f(z)$ can vanish only at the λ_n's with $n \neq 0$. For if $f(z)$ were zero for some other value, $z = \gamma$, say, then the function

$$\frac{zf(z)}{z - \gamma}$$

would also belong to $\boldsymbol{P}$ and would vanish at every λ_n. The system $\{e^{i\lambda_n t}\}$ would then be incomplete in $L^2[-\pi, \pi]$, contrary to hypothesis. By Hadamard's factorization theorem

$$f(z) = e^{Az} \prod_{n=1}^{\infty} \left(1 - \frac{z^2}{\lambda_n^2}\right).$$

Since $f(z)$ and the canonical product are both even, $A = 0$ and hence

$$f(z) = \prod_{n=1}^{\infty} \left(1 - \frac{z^2}{\lambda_n^2}\right). \blacksquare$$

Returning now to Example 3, we write

$$G(z) = z \prod_{n=1}^{\infty} \left(1 - \frac{z^2}{\lambda_n^2}\right).$$

For each integer n, the entire function $G_n(z)$ defined by

$$G_n(z) = \frac{G(z)}{G'(\lambda_n)(z - \lambda_n)}$$

belongs to the Paley–Wiener space. Assertion: g_n is the inverse Fourier transform of G_n, i.e., for almost all $t \in [-\pi, \pi]$,

$$g_n(t) = \int_{-\infty}^{\infty} G_n(x) e^{ixt}\, dx,$$

where the integral is understood to be a *limit in the mean* (in the L^2 sense). Call the right side $h_n(t)$. By the Paley–Wiener theorem, $h_n(t) = 0$ almost everywhere outside $[-\pi, \pi]$. By the Fourier inversion formula,

$$G_n(x) = \frac{1}{2\pi} \int_{-\pi}^{\pi} h_n(t) e^{-ixt}\, dt,$$

and therefore

$$\frac{1}{2\pi} \int_{-\pi}^{\pi} h_n(t) e^{-i\lambda_m t}\, dt = \delta_{nm}.$$

Since the system $\{e^{i\lambda_n t}\}$ has a *unique* biorthogonal sequence, we must have $g_n = h_n$ for every n, as asserted.

By means of the Fourier isometry, we can transform the entire preceding discussion into the Paley–Wiener space. The exponentials $e^{i\lambda_n t}$ are transformed into the reproducing functions

$$K_n(z) = \frac{\sin \pi(z - \lambda_n)}{\pi(z - \lambda_n)},$$

$g_n(t)$ is transformed into $G_n(z)$, while the moment problem itself becomes

$$f(\lambda_n) = c_n \quad (n = 0, \pm 1, \pm 2, \ldots),$$

since $f(\lambda_n) = (f, K_n)$. Here $\{c_n\} \in l^2$ and $f \in \boldsymbol{P}$ is to be found. The solution is immediate: $f(z)$ is given by the norm-convergent series

$$f(z) = \sum_{n=-\infty}^{\infty} \frac{c_n G(z)}{G'(\lambda_n)(z - \lambda_n)}. \tag{3}$$

Moreover, since the expansion

$$f = \sum_{n=-\infty}^{\infty} (f, K_n) G_n$$

is valid for every function f belonging to $\boldsymbol{P}$ and $\{G_n\}$ is a Riesz basis for $\boldsymbol{P}$, it follows that

$$\sum |f(\lambda_n)|^2 < \infty \quad \text{for all } f.$$

Thus, (3), with $\{c_n\}$ in l^2, represents the *most general* function in $\boldsymbol{P}$.

Formula (3) is a simple example of a "generalized" **Lagrange interpolation formula** for an entire function assuming the values c_n at the points λ_n. We have obtained it under the assumption that $\{\lambda_n\}$ is a symmetric sequence of real numbers for which the corresponding system of exponentials $\{e^{i\lambda_n t}\}$ constitutes a Riesz basis for $L^2[-\pi, \pi]$. In particular, the formula is valid whenever $\{\lambda_n\}$ is real, symmetric, and

$$|\lambda_n - n| \leqq L < \tfrac{1}{4}, \quad n = 0, \pm 1, \pm 2, \ldots,$$

by virtue of Theorem 1.14. If we choose $\lambda_n = n$, then $G(z) = (\sin \pi z)/\pi$, and (3) reduces to the "cardinal series" for $f(z)$.

The preceding examples indicate that there can be no simple solution to the general moment problem $(f, f_n) = c_n$, $n = 1, 2, 3, \ldots$. The following theorem provides a criterion for a solution which, although both necessary and sufficient, is in practice difficult to apply. Nevertheless, in the absence of additional information about the f_n's, it is frequently of use.

Theorem 2. *Let $\{f_1, f_2, f_3, \ldots\}$ be a sequence of vectors belonging to a Hilbert space H and $\{c_1, c_2, c_3, \ldots\}$ a sequence of scalars. In order that the equations*

$$(f, f_n) = c_n \quad (n = 1, 2, 3, \ldots)$$

shall admit at least one solution $f \in H$ for which $\|f\| \leqq M$, it is necessary and sufficient that

$$\left|\sum a_n \bar{c}_n\right| \leqq M \left\|\sum a_n f_n\right\| \tag{4}$$

for every finite sequence of scalars $\{a_n\}$.

Proof. The necessity is trivial: if $\{a_n\}$ is an arbitrary finite sequence of scalars, then

$$\left|\sum a_n \bar{c}_n\right| = \left|\left(f, \sum a_n f_n\right)\right| \leqq M \left\|\sum a_n f_n\right\|.$$

For the sufficiency, let Y be the subspace spanned by the elements f_1, f_2, $f_3, \dots$ and define μ on Y by setting

$$\mu\left(\sum a_n f_n\right) = \sum a_n \bar{c}_n$$

for every finite sequence of scalars $\{a_n\}$. Condition (4) guarantees that μ is defined unambiguously as a bounded linear functional on Y, with norm not exceeding M. It follows that we may extend μ to all of H so that $|\mu(f)| \leqq M\|f\|$. (Notice that the possibility of extending μ in this way need not rely on the Hahn-Banach theorem: first extend μ to the closure of Y "by continuity" and then define $\mu(f) = 0$ for $f \in Y^{\perp}$.)

By the Riesz representation theorem, there is a vector $f \in H$ such that $\mu(g) = (g, f)$ for every $g \in H$. It follows that $(f, f_n) = c_n$ $(n = 1, 2, 3, \dots)$ and $\|f\| = \|\mu\| \leqq M$. ■

PROBLEMS

1. Let X be an n-dimensional vector space, X' its algebraic dual, and $\{x_1, \dots, x_n\}$ and $\{f_1, \dots, f_n\}$ linearly independent subsets of X and X', respectively. Prove that the generalized Gram determinant

$$\det(f_i(x_j))_{i,j=1}^{n}$$

is not zero.

2. Let $\{f_1, f_2, f_3, \dots\}$ be a linearly independent set of vectors belonging to a Hilbert space H, and suppose that the system

$$(f, f_n) = c_n \quad (n = 1, 2, 3, \dots)$$

admits at least one solution. Let h be the (unique) solution of minimal norm. Let h_n $(n = 1, 2, 3, \dots)$ be the (unique) solution of minimal norm of the system

$$(f, f_i) = c_i \quad (i = 1, \dots, n).$$

Show that $h_n \to h$ in norm.

3. Let $\{f_1, f_2, f_3, \dots\}$ be a basis for a Hilbert space H and M its corresponding moment space. Show that $\{f_n\}$ is a *Bessel basis* if and only if $M \supset l^2$ and a *Hilbert basis* if and only if $M \subset l^2$. (For the definitions of Bessel and Hilbert bases, see Problem 4, Section 1.8.)

4. Prove that if $f(t) \in L^2[0, 1]$ and $c_n = \int_0^1 t^n f(t)\,dt$, $n = 0, 1, 2, \dots$, then $\{c_n\} \in l^2$. (*Hint*: Show that the series $\sum_{n=0}^{\infty} b_n c_n$ is convergent whenever $\{b_n\} \in l^2$.)

5. Let $\{\lambda_n\}$ be a symmetric sequence of real or complex numbers such that the system of exponentials $\{e^{i\lambda_n t}\}$ forms a Riesz basis for $L^2[-\pi, \pi]$. Show that the canonical product

$$\prod_{n=1}^{\infty}\left(1-\frac{z^2}{\lambda_n^2}\right)$$

is an entire function of exponential type π.

6. Let $x_1, \ldots, x_n$ be n linearly independent vectors from a normed vector space X and let $c_1, \ldots, c_n$ be arbitrary scalars, *not all zero*. Show that there exists an element $f \in X^*$ such that

$$f(x_i) = c_i \quad (i = 1, \ldots, n). \tag{5}$$

7. With $\{x_1, \ldots, x_n\}$ and $\{c_1, \ldots, c_n\}$ as in Problem 6, define $\lambda = \lambda(c_1, \ldots, c_n)$ by putting

$$\frac{1}{\lambda} = \inf \|\xi_1 x_1 + \cdots + \xi_n x_n\|,$$

where the infimum is over all scalars $\xi_1, \ldots, \xi_n$ for which

$$c_1\xi_1 + \cdots + c_n\xi_n = 1.$$

(a) Show that the infimum is attained.
(b) Show that the system (5) in Problem 6 has at least one solution $f \in X^*$ for which $\|f\| \leqq L$ if and only if $L \geqq \lambda$.
(c) Show that λ is the *minimum* of the norms of all elements $f \in X^*$ that satisfy (5).

2 BESSEL SEQUENCES AND RIESZ–FISCHER SEQUENCES

Bessel's inequality and the Riesz–Fischer theorem together embody one of the fundamental characteristics of an orthonormal basis: the moment space is l^2. Because of their intrinsic importance, these properties have been abstracted.

Definition. *A sequence $\{f_1, f_2, f_3, \ldots\}$ of vectors belonging to a Hilbert space H is said to be a* ***Bessel sequence*** *if*

$$\sum_{n=1}^{\infty} |(f, f_n)|^2 < \infty$$

for every element $f \in H$. The sequence is called a ***Riesz–Fischer sequence*** *if the moment problem*

$$(f, f_n) = c_n \quad (n = 1, 2, 3, \ldots) \tag{1}$$

admits at least one solution $f \in H$ whenever $\{c_n\} \in l^2$.

Thus, $\{f_n\}$ is a Bessel sequence whenever its moment space is a subset of l^2, while it is a Riesz–Fischer sequence if its moment space contains l^2.

Proposition 2. *Let $\{f_n\}$ be a sequence of vectors belonging to a Hilbert space H. If $\{f_n\}$ is a Bessel sequence, then there exists a constant M such that*

$$\sum_{n=1}^{\infty} |(f, f_n)|^2 \leqq M\|f\|^2$$

for every f in H. If $\{f_n\}$ is a Riesz–Fischer sequence, then there exists a constant m such that (1) *has at least one solution f satisfying*

$$\|f\|^2 \leqq \frac{1}{m} \sum_{n=1}^{\infty} |c_n|^2,$$

provided $\{c_n\} \in l^2$.

The proof of the proposition is left to the reader (see Problem 3). The numbers m and M appearing in the proposition are called *bounds* for the sequence.

We state the following corollary separately in view of its importance for applications.

Corollary. *If $\{f_n\}$ is a Riesz–Fischer sequence in a Hilbert space, then the unit ball of l^2 can be interpolated in a uniformly bounded way.*

The next theorem provides a fundamental characterization of Bessel sequences and Riesz–Fischer sequences in a Hilbert space.

Theorem 3. *Let $\{f_1, f_2, f_3, \ldots\}$ be a sequence of vectors belonging to a Hilbert space H. Then*

(i) *$\{f_1, f_2, f_3, \ldots\}$ is a Bessel sequence with bound M if and only if the inequality*

$$\left\|\sum c_n f_n\right\|^2 \leqq M \sum |c_n|^2$$

holds for every finite sequence of scalars $\{c_n\}$;

(ii) *$\{f_1, f_2, f_3, \ldots\}$ is a Riesz–Fischer sequence with bound m if and only if the inequality*

$$m \sum |c_n|^2 \leqq \left\|\sum c_n f_n\right\|^2$$

holds for every finite sequence of scalars $\{c_n\}$.

Proof. (i) For the necessity, let $\{c_n\}$ be an arbitrary finite sequence of scalars and put $f = \sum c_n f_n$. Then

$$\|f\|^4 = |(f, f)|^2 = \left|\sum \bar{c}_n (f, f_n)\right|^2$$
$$\leqq \sum |c_n|^2 \sum |(f, f_n)|^2 \leqq M\|f\|^2 \sum |c_n|^2,$$

by Proposition 2. Dividing by $\|f\|^2$, we find

$$\|f\|^2 \leqq M \sum |c_n|^2.$$

For the sufficiency, observe first that

$$\left\|\sum_{n=1}^{\infty} c_n f_n\right\|^2 \leqq M \sum_{n=1}^{\infty} |c_n|^2$$

whenever $\{c_n\} \in l^2$. If f is a fixed but arbitrary element of H, then for every square-summable sequence $\{c_n\}$ we have

$$\left|\sum_{n=1}^{\infty} c_n (f, f_n)\right|^2 = \left|\left(f, \sum_{n=1}^{\infty} \bar{c}_n f_n\right)\right|^2$$
$$\leqq \|f\|^2 \cdot \left\|\sum_{n=1}^{\infty} \bar{c}_n f_n\right\|^2 \leqq M\|f\|^2 \sum_{n=1}^{\infty} |c_n|^2.$$

It follows that

$$\sum_{n=1}^{\infty} |(f, f_n)|^2 \leqq M\|f\|^2,$$

by the converse to Hölder's inequality.

(ii) For the necessity, fix an arbitrary finite sequence of scalars $\{c_1, \ldots, c_N\}$ and choose $a_1, a_2, a_3, \ldots$ so that (1) $a_n = 0$ if $n > N$, (2) $\sum_{n=1}^{N} |a_n|^2 = 1$, and (3) $|\sum_{n=1}^{N} \bar{a}_n c_n|^2 = \sum_{n=1}^{N} |c_n|^2$.

By Proposition 2, there is a constant m (independent of N and $c_1, \ldots, c_N$) such that the moment problem

$$(f, f_n) = a_n \quad (n = 1, 2, 3, \ldots) \tag{3}$$

has at least one solution f with $\|f\|^2 \leqq 1/m$. Applying Theorem 2, we find that

$$\sum_{n=1}^{N} |c_n|^2 = \left|\sum_{n=1}^{N} \bar{a}_n c_n\right|^2 \leqq \frac{1}{m} \left\|\sum_{n=1}^{N} c_n f_n\right\|^2,$$

and the result follows.

As for the sufficiency, it is enough to show that (3) has a solution f with $\|f\|^2 \leqq 1/m$ whenever $\sum_{n=1}^{\infty} |a_n|^2 \leqq 1$. Suppose then that $\{a_n\}$ belongs to the unit ball of l^2 and let $c_1, \ldots, c_N$ be arbitrary scalars. Since

$$\left|\sum_{n=1}^{N} \overline{a}_n c_n\right|^2 \leqq \sum_{n=1}^{N} |c_n|^2 \leqq \frac{1}{m} \left\|\sum_{n=1}^{N} c_n f_n\right\|,$$

the result is, once again, a consequence of Theorem 2. ■

The criteria of Theorem 3 can be phrased more succinctly in the language of operators. Thus, $\{f_1, f_2, f_3, \ldots\}$ is a Bessel sequence if and only if, given an arbitrary orthonormal basis $\{e_1, e_2, e_3, \ldots\}$ for H, there exists a bounded linear operator $T : H \to H$ such that

$$Te_n = f_n \quad \text{for} \quad n = 1, 2, 3, \ldots;$$

it is a Riesz–Fischer sequence if and only if there exists a bounded linear operator $S : H \to H$ such that

$$Sf_n = e_n \quad \text{for} \quad n = 1, 2, 3, \ldots.$$

An equivalent formulation in terms of the Gram matrix

$$A = ((f_i, f_j))_{i,j=1}^{\infty}$$

is also useful. In this setting, $\{f_1, f_2, f_3, \ldots\}$ is a Bessel sequence with bound M if and only if A defines a bounded (self-adjoint) operator on l^2, with norm not exceeding M. Reason: A is bounded by M if and only if the associated quadratic form

$$\sum_{i,j=1}^{\infty} (f_i, f_j) c_i \overline{c}_j$$

is bounded by M, and this form equals $\|\sum_{i=1}^{\infty} c_i f_i\|^2$. Similarly, $\{f_1, f_2, f_3, \ldots\}$ is a Riesz–Fischer sequence with bound m if and only if the "sections" of A are *bounded from below by m*. This means that

$$m\|c\| \leqq \|A_n c\|$$

for all n-tuples $c = (c_1, \ldots, c_n)$ and all n. Here the norms are Euclidean norms, and A_n is the $n \times n$ matrix $((f_i, f_j))_{i,j=1}^{n}$. The reader will verify that this is equivalent to

$$m\|c\|^2 \leqq (A_n c, c)$$

for all n and c; it is also equivalent to $\|A_n^{-1}\| \leqq 1/m$ for all n.

Example 1. The sequence of powers $\{1, t, t^2, \ldots\}$ forms a Bessel sequence in $L^2[0, 1]$. In fact, if $f \in L^2[0, 1]$ and $c_n = \int_0^1 t^n f(t)\, dt$ $(n = 0, 1, 2, \ldots)$, then

$$\sum_{n=0}^{\infty} |c_n|^2 \leqq \pi \int_0^1 |f(t)|^2\, dt.$$

The constant π is the best possible. Reason: The corresponding Gram matrix

$$A = ((t^i, t^j))_{i,j=0}^{\infty}$$

is the "Hilbert matrix" (the i, jth entry is $1/(i + j + 1)$), and so defines a bounded linear operator on l^2 of norm π (see Hardy, Littlewood, and Pólya [1952]; also see Problem 5).

Suppose now that H is a functional Hilbert space, over a set S say, with reproducing kernel K, and let $\{\lambda_1, \lambda_2, \lambda_3, \ldots\}$ be a sequence of points in S. We shall take for the f_n's the normalized kernel functions $K_n/\|K_n\|$, where $K_n(z) = K(z, \lambda_n)$. Then $\{f_n\}$ is a Bessel sequence if

$$\sum_{n=1}^{\infty} \frac{|f(\lambda_n)|^2}{K(\lambda_n, \lambda_n)} < \infty \tag{4}$$

for every function $f \in H$; it is a Riesz–Fischer sequence if the "weighted" interpolation problem

$$\frac{f(\lambda_n)}{\sqrt{K(\lambda_n, \lambda_n)}} = c_n \quad (n = 1, 2, 3, \ldots) \tag{5}$$

has a solution whenever $\{c_n\} \in l^2$.

Example 2. As an illustration, let us take H to be the Hardy space $\boldsymbol{H}^2$ and $\{\lambda_1, \lambda_2, \lambda_3, \ldots\}$ a sequence of distinct points in the open unit disk $|z| < 1$. Since the reproducing kernel of $\boldsymbol{H}^2$ (the Szegö kernel) is given by $K(z, w) = 1/(1 - z\overline{w})$, conditions (4) and (5) become, respectively,

$$\sum_{n=1}^{\infty} |f(\lambda_n)|^2 (1 - |\lambda_n|^2) < \infty \tag{6}$$

and

$$f(\lambda_n)\sqrt{1 - |\lambda_n|^2} = c_n \quad (n = 1, 2, 3, \ldots). \tag{7}$$

We are going to show that (6) holds for every $f \in \boldsymbol{H}^2$ provided that λ_n approaches the boundary of the unit disk *exponentially*. This means that there exists a constant $\alpha < 1$ such that

$$1 - |\lambda_{n+1}| \leqq \alpha(1 - |\lambda_n|), \quad n = 1, 2, 3, \ldots.$$

It is enough to show that the Gram matrix $A = (a_{ij})_{i,j=1}^{\infty}$, where

$$a_{ij} = \left(\frac{K_i}{\|K_i\|}, \frac{K_j}{\|K_j\|} \right) = \frac{\sqrt{1 - |\lambda_i|^2}\sqrt{1 - |\lambda_j|^2}}{1 - \overline{\lambda}_i \lambda_j}$$

defines a bounded operator on l^2. For this purpose the following simple form of **Schur's lemma** will be needed.

Lemma 1. *Let $A = (a_{ij})$ be an infinite Hermitian matrix ($a_{ij} = \bar{a}_{ji}$), and suppose that for some constant M*

$$\sum_{i=1}^{\infty} |a_{ij}| \leqq M \quad (j = 1, 2, 3, \ldots).$$

Then A defines a bounded linear operator on l^2 of norm not exceeding M.

Proof. It is sufficient to show that the associated quadratic form (Ax, x) is bounded by M on l^2, i.e., that

$$\left| \sum_{i,j=1}^{\infty} a_{ij} x_i \bar{x}_j \right| \leqq M \sum_{i=1}^{\infty} |x_i|^2 \tag{8}$$

for every square-summable sequence $x = \{x_i\}$. If we write

$$|a_{ij} x_i \bar{x}_j| = \left(|a_{ij}|^{1/2} |x_i| \right) \left(|a_{ij}|^{1/2} |x_j| \right)$$

and apply the Cauchy-Schwarz inequality twice, then (8) follows readily. ∎

Let us return now to the Gram matrix under consideration. It follows by the assumption on $\{\lambda_n\}$ that

$$1 - |\lambda_{i+n}| \leqq \alpha^n (1 - |\lambda_i|)$$

and

$$|1 - \bar{\lambda}_i \lambda_j| \geqq 1 - |\lambda_{\min(i,j)}|.$$

Accordingly, for each j,

$$\begin{aligned}
\sum_{i=1}^{\infty} |a_{ij}| &= \sum_{i=1}^{j} |a_{ij}| + \sum_{i=j+1}^{\infty} |a_{ij}| \\
&\leqq 2 \sum_{i=1}^{j} \left(\frac{1 - |\lambda_j|}{1 - |\lambda_i|} \right)^{1/2} + 2 \sum_{i=j+1}^{\infty} \left(\frac{1 - |\lambda_i|}{1 - |\lambda_j|} \right)^{1/2} \\
&\leqq 2 \sum_{i=1}^{j} \alpha^{(j-i)/2} + 2 \sum_{i=j+1}^{\infty} \alpha^{(i-j)/2} \\
&< 2 \left(\frac{1}{1 - \alpha^{1/2}} + \frac{\alpha^{1/2}}{1 - \alpha^{1/2}} \right) = 2 \left(\frac{1 + \alpha^{1/2}}{1 - \alpha^{1/2}} \right).
\end{aligned}$$

This establishes the Bessel condition (6) for all f in $\boldsymbol{H}^2$.

Shapiro and Shields [1961] have shown that if λ_n approaches the boundary of the unit disk exponentially, then the validity of (6) for every $f \in \boldsymbol{H}^2$ is equivalent to the solvability of (7) for every sequence $\{c_n\} \in l^2$.

There is a growing literature on interpolation problems in a wide variety of Banach spaces of analytic functions. For additional references, see the *Notes* (see, especially, Duren [1970], in which an account is given of Carleson's proof of the "corona theorem").

PROBLEMS

1. We shall say that a sequence of elements from a normed vector space is *uniformly minimal* if for some $\varepsilon > 0$ the distance from each element of the sequence to the closed linear span of the others is at least ε. Show that every Riesz–Fischer sequence is uniformly minimal. Is the converse true?

2. Let $\{f_n\}$ and $\{g_n\}$ be complete biorthogonal sequences belonging to a Hilbert space H. Show that $\{f_n\}$ is a Bessel sequence if and only if $\{g_n\}$ is a Riesz–Fischer sequence. Is completeness necessary?

3. Prove Proposition 2. (*Hint*: If $\{f_n\}$ is a Bessel sequence, then the mapping $T: H \to l^2$ defined by $Tf = \{(f, f_n)\}$ has a closed graph. If $\{f_n\}$ is a Riesz–Fischer sequence, then there is a natural mapping

$$S: l^2 \to H/[f_1, f_2, f_3, \ldots]^{\perp},$$

and this mapping has a closed graph.)

4. Prove that a sequence of vectors $\{f_n\}$ in a Hilbert space H is a Riesz–Fischer sequence if and only if the eigenvalues of the sections $A_n = ((f_i, f_j))_{i,j=1}^n$, $n = 1, 2, 3, \ldots$, are bounded away from zero.

5. (a) Prove **Hilbert's inequality**: if $c_n \geqq 0$ $(n = 0, 1, 2, \ldots)$, then

$$\sum_{m,n=0}^{\infty} \frac{c_m c_n}{m+n+1} \leqq \pi \sum_{n=0}^{\infty} |c_n|^2.$$

Conclude that the Hilbert matrix defines a bounded linear operator on l^2 of norm not exceeding π. (*Hint*: Put $f(z) = \sum_{n=0}^{N} c_n z^n$. Using Cauchy's theorem, show that

$$\int_{-1}^{1} |f(x)|^2\, dx \leqq \frac{1}{2} \int_{-\pi}^{\pi} |f(e^{i\theta})|^2\, d\theta.$$

This is the **Fejér–Riesz inequality**.)

(b) By choosing $f(x) = (1-x)^{\varepsilon - 1/2}$, $\varepsilon > 0$, show that the constant π appearing in Example 1 is the best possible. Conclude that the norm of the Hilbert matrix must, in fact, be *equal* to π.

6. Show that the Hilbert matrix is not invertible.

7. Show that the system (7) is solvable (with $f \in \boldsymbol{H}^2$) for every sequence $\{c_n\} \in l^2$ if

$$\lambda_n = 1 - 2^{-n} \quad (n = 1, 2, 3, \ldots).$$

(*Hint*: The Gram matrix $A = (a_{ij})$, where $a_{ij} = (K_i/\|K_i\|, K_j/\|K_j\|)$, is invertible.)

8. (Interpolation in $\boldsymbol{B}_\tau$) Let T be the mapping of $\boldsymbol{B}_\tau$ to l^∞ defined by

$$Tf = \{f(n)\}_{-\infty}^{\infty}.$$

(a) Show that T maps $\boldsymbol{B}_\tau$ onto l^∞ if $\tau > \pi$.
(*Hint*: If $\varepsilon > 0$ and if $\{c_n\} \in l^\infty$, then the series

$$f(z) = \sum_{n=-\infty}^{\infty} c_n \frac{\sin \pi(z-n)}{\pi(z-n)} \frac{\sin \varepsilon(z-n)}{\varepsilon(z-n)}$$

converges in $\boldsymbol{B}_\tau$.)

(b) Show that T does not map $\boldsymbol{B}_\pi$ *onto* l^∞. (*Hint*: The sequence $\{c_n\}$ defined by

$$c_n = \begin{cases} 0 & \text{if } \; n \leqq 0 \\ (-1)^n & \text{if } \; n > 0 \end{cases}$$

cannot be interpolated (cf. Problem 4, Section 2.5, Part Two).)

3 APPLICATIONS TO SYSTEMS OF COMPLEX EXPONENTIALS

In Section 2 we considered Bessel sequences and Riesz–Fischer sequences in an abstract Hilbert space, and we showed that these properties are equivalent to a number of others. In the present section we shall apply these results to systems of complex exponentials $\{e^{i\lambda_n t}\}$ in $L^2[-A, A]$. As a result, we shall obtain added information about closure and the completeness radius.

Theorem 4. *If $\{\lambda_n\}$ is a separated sequence of real numbers, then the system $\{e^{i\lambda_n t}\}$ forms a Bessel sequence in $L^2[-A, A]$ for every positive number A. If $\sum |c_n|^2 < \infty$, then the series*

$$\sum_n c_n e^{i\lambda_n t}$$

converges in the mean to an element of $L^2[-A, A]$.

Proof. If $\phi \in L^2[-A, A]$, then the inner product

$$a_n = (\phi(t), e^{i\lambda_n t})$$

is just the value $f(\lambda_n)$ of the entire function

$$f(z) = \frac{1}{2A} \int_{-A}^{A} \phi(t) e^{-izt}\, dt.$$

Since $f(z)$ is of exponential type (at most A) and belongs to L^2 along the real axis, we have

$$\sum_n |a_n|^2 = \sum_n |f(\lambda_n)|^2 < \infty$$

(by Theorem 2.17). This proves the first statement. The second follows from the first by Theorem 3. ■

If the separation of the λ_n's is great enough, then $\{e^{i\lambda_n t}\}$ is also a Riesz–Fischer sequence.

Theorem 5. *If $\{\lambda_n\}$ is a separated sequence of real numbers such that*

$$\lambda_{n+1} - \lambda_n \geqq \gamma > \frac{\pi}{A} \quad (n = 0, \pm 1, \pm 2, \ldots),$$

then the system $\{e^{i\lambda_n t}\}$ is a Riesz–Fischer sequence in $L^2[-A, A]$.

Proof. Without loss of generality, we may assume that $A = \pi$, so that $\gamma > 1$. It is to be shown that for all finite sequences $\{c_n\}$ of scalars,

$$m \sum |c_n|^2 \leqq \left\| \sum c_n e^{i\lambda_n t} \right\|^2,$$

where

$$m = \frac{2}{\pi}\left(1 - \frac{1}{\gamma^2}\right) > 0.$$

Let $\{c_n\}$ be such a sequence and put

$$f(t) = \sum c_n e^{i\lambda_n t}.$$

If $k(t)$ is any function integrable over the real axis and

$$K(x) = \int_{-\infty}^{\infty} k(t) e^{ixt}\, dt, \qquad -\infty < x < \infty,$$

then

$$\int_{-\infty}^{\infty} k(t) |f(t)|^2\, dt = \sum_{m,n} K(\lambda_m - \lambda_n) c_m \overline{c}_n.$$

If we choose

$$k(t) = \begin{cases} \cos \frac{1}{2} t & \text{if } \ |t| \leqq \pi, \\ 0 & \text{if } \ |t| > \pi, \end{cases}$$

then

$$K(x) = \frac{4 \cos \pi x}{1 - 4x^2},$$

and

$$\int_{-\pi}^{\pi} |f(t)|^2\,dt \geqq \sum_{m,n} K(\lambda_m - \lambda_n) c_m \overline{c}_n.$$

Let S_1 denote that part of the sum for which $m = n$ and S_2 the remaining part. Clearly,

$$S_1 = 4 \sum_n |c_n|^2.$$

Since $K(x)$ is even and since $2|c_m \overline{c}_n| \leqq |c_m|^2 + |c_n|^2$, there is a constant θ such that $|\theta| \leqq 1$ and

$$S_2 = \theta \sum_{m \neq n} \frac{|c_m|^2 + |c_n|^2}{2} |K(\lambda_m - \lambda_n)| = \theta \sum_n |c_n|^2 {\sum_m}' |K(\lambda_m - \lambda_n)|$$

(the prime indicates omission of the term $m = n$). Since

$$|\lambda_m - \lambda_n| \geqq |m - n|\gamma > 1$$

for $m \neq n$, we have

$$\begin{aligned} {\sum_m}' |K(\lambda_m - \lambda_n)| &\leqq {\sum_m}' \frac{4}{4(m-n)^2\gamma^2 - 1} < \frac{8}{\gamma^2} \sum_{r=1}^{\infty} \frac{1}{4r^2 - 1} \\ &= \frac{4}{\gamma^2} \sum_{r=1}^{\infty} \left(\frac{1}{2r-1} - \frac{1}{2r+1} \right) = \frac{4}{\gamma^2}. \end{aligned}$$

Hence

$$\int_{-\pi}^{\pi} |f(t)|^2\,dt \geqq \left(4 - \frac{4}{\gamma^2}\right) \sum |c_n|^2,$$

and the theorem follows. ∎

The theorem is sharp in the sense that if $\gamma = \pi/A$, then the conclusion no longer holds. The following example is by now a familiar one.

Example. The system of exponentials $\{e^{\pm i(n-1/4)t} : n = 1, 2, 3, \ldots\}$ is not a Riesz–Fischer sequence in $L^2[-\pi, \pi]$.

Write $\lambda_n = n - \frac{1}{4}$ and $\lambda_{-n} = -\lambda_n$ $(n = 1, 2, 3, \ldots)$. Let $g(z) = \sum_{n=0}^{\infty} a_n z^n$ be analytic for $|z| < 1$ and define

$$f(t) = e^{i\lambda t} g(re^{it}) + e^{-i\lambda t} g(re^{-it}),$$

where $0 < r < 1$ and $\lambda > 0$. Then

$$f(t) = \sum_{n=0}^{\infty} a_n r^n e^{i(n+\lambda)t} + \sum_{n=0}^{\infty} a_n r^n e^{-i(n+\lambda)t},$$

and both series converge uniformly in t for fixed values of r and λ. If we take $\lambda = \frac{3}{4}$, then $f(t)$ is of the form

$$f(t) = \sum c_n e^{i\lambda_n t}$$

with $\lambda_{n+1} - \lambda_n \geqq 1$.

If the system of exponentials $\{e^{i\lambda_n t}\}$ were in fact a Riesz–Fischer sequence in $L^2[-\pi, \pi]$, then we could find a positive constant m such that

$$m \sum |c_n|^2 \leqq \left\| \sum c_n e^{i\lambda_n t} \right\|^2.$$

Since

$$\sum |c_n|^2 = 2 \sum_{n=0}^{\infty} |a_n|^2 r^{2n} = \frac{1}{\pi} \int_{-\pi}^{\pi} |g(re^{it})|^2 \, dt,$$

this would imply

$$m \int_{-\pi}^{\pi} |g(re^{it})|^2 \, dt \leqq \frac{1}{2} \int_{-\pi}^{\pi} |f(t)|^2 \, dt \tag{1}$$

for *every* r $(0 < r < 1)$.

Take

$$g(z) = (1 + z)^{-1/2}$$

(the principal value is to be chosen). We leave it to the reader to show that for this choice of g, the integral on the right in (1) remains bounded as $r \to 1$, while that on the left is asymptotically $-2\pi \log(1 - r)$. But this is absurd, and hence $\{e^{i\lambda_n t}\}$ is not a Riesz–Fischer sequence in $L^2[-\pi, \pi]$.

The following result is an almost immediate consequence of Theorem 5.

Theorem 6. *If $\{\lambda_n\}$ is a separated sequence of real numbers such that*

$$\lambda_{n+1} - \lambda_n \geqq \gamma > \frac{\pi}{A} \qquad (n = 0, \pm 1, \pm 2, \ldots),$$

then the system $\{e^{i\lambda_n t}\}$ has infinite deficiency in $L^2[-A, A]$.

Proof. Let N be a fixed but arbitrary positive integer. If K is large enough, then we can replace

$$\lambda_0, \lambda_1, \ldots, \lambda_K$$

by

$$\mu_0, \mu_1, \ldots, \mu_{K+N+1},$$

so that the resulting sequence, relabeled $\{\mu_n\}$, satisfies

$$\inf_n (\mu_{n+1} - \mu_n) > \frac{\pi}{A}.$$

By Theorem 5 there is a function $\phi \in L^2[-A, A]$ such that

$$\int_{-A}^{A} \phi(t) e^{-i\mu_n t}\, dt = \begin{cases} 1 & \text{if} \quad n = 0, \\ 0 & \text{if} \quad n \neq 0. \end{cases}$$

Thus the system $\{e^{i\mu_n t}: n \neq 0\}$ is incomplete in $L^2[-A, A]$, and we conclude by Theorem 3.7 that the deficiency of $\{e^{i\lambda_n t}\}$ in $L^2[-A, A]$ is at least N. Since N was arbitrary, the result follows. ■

Corollary. *Every sequence* $\{\lambda_n\}_{-\infty}^{\infty}$ *of real numbers for which*

$$\lambda_{n+1} - \lambda_n \to \infty \quad as \quad n \to \pm\infty$$

has completeness radius zero.

PROBLEMS

1. Let $\{\lambda_n\}$ be a sequence of real numbers such that $\lambda_{n+1} - \lambda_n \geqq 1$ for all n. Theorems 3 and 4 imply that there is a constant M such that

$$\left\| \sum c_n e^{i\lambda_n t} \right\|^2 \leqq M \sum |c_n|^2$$

whenever $\{c_n\} \in l^2$. (The norm is in $L^2[-\pi, \pi]$.) Show that M must be at least 2.

2. Give an "elementary" proof of Theorem 4 by using the same method used to prove Theorem 5.

3. Prove that the system $\{e^{\pm i(n+1/4)t}: n = 1, 2, 3, \ldots\}$ is not a Riesz–Fischer sequence in $L^2[-\pi, \pi]$.

4. Every basis $\{e^{i\lambda_n t}\}$ for $L^2[-A, A]$ such that the exponents λ_n are all *real* is a Hilbert basis. (*Hint*: Show that $\{\lambda_n\}$ is separated. For the definition of a Hilbert basis, see Problem 4, Section 1.8.)

4 THE MOMENT SPACE AND ITS RELATION TO EQUIVALENT SEQUENCES

Let $\{f_1, f_2, f_3, \ldots\}$ be a sequence of vectors belonging to a separable Hilbert space. In what ways are the properties of the sequence reflected in its moment space? We have seen, for example, that if the sequence is a Riesz basis, then its moment space coincides with l^2. The converse, of course, is false: the removal of a single vector from a Riesz basis leaves an incomplete set whose moment space is again l^2. The goal of the present section is to show that the converse is true provided that $\{f_1, f_2, f_3, \ldots\}$ is complete: a complete sequence of vectors belonging to a Hilbert space is a Riesz basis if and only if its moment space coincides with l^2. We begin by showing that the moment space characterizes complete sequences up to "equivalence."

Two sequences of vectors $\{f_1, f_2, f_3, \ldots\}$ and $\{g_1, g_2, g_3, \ldots\}$ belonging to a Hilbert space H are said to be **equivalent** if there exists a bounded invertible operator $T: H \to H$ such that $Tf_n = g_n$ $(n = 1, 2, 3, \ldots)$.

Theorem 7. *Two complete sequences of vectors belonging to a separable Hilbert space are equivalent if and only if they have the same moment space.*

Proof. Let $\{f_1, f_2, f_3, \ldots\}$ and $\{g_1, g_2, g_3, \ldots\}$ be complete sequences in H and suppose that there exists a bounded invertible operator T on H such that $Tf_n = g_n$ $(n = 1, 2, 3, \ldots)$. It is to be shown that the system of equations

$$(f, f_n) = (g, g_n) \quad (n = 1, 2, 3, \ldots) \tag{1}$$

admits a (unique) solution f, whenever g is given, and a (unique) solution g, whenever f is given. Evidently this is true if

$$f = T^* g,$$

and the condition is necessary.

Suppose now that $\{f_1, f_2, f_3, \ldots\}$ and $\{g_1, g_2, g_3, \ldots\}$ have the same moment space. Then for each g in H there corresponds a unique vector f such that (1) holds. We are going to show that this implies the existence of a bounded linear operator $T: H \to H$ such that

$$Tf_n = g_n \quad (n = 1, 2, 3, \ldots). \tag{2}$$

Let Y be the linear span of the f_n's. If $f \in Y$, with $f = \sum_{n=1}^{N} c_n f_n$, define Tf by putting

$$Tf = \sum_{n=1}^{N} c_n g_n.$$

We first show that T is well-defined. If it were not, then we could find scalars $c_1, \ldots, c_N$ such that

$$\sum_{n=1}^{N} c_n f_n = 0 \quad \text{and} \quad \sum_{n=1}^{N} c_n g_n \neq 0.$$

Choosing $g = \sum_{n=1}^{N} c_n g_n$ and then taking f to be the corresponding solution to (1), we find

$$\begin{aligned} 0 &= \left(f, \sum_{n=1}^{N} c_n f_n \right) = \sum_{n=1}^{N} \overline{c}_n (f, f_n) = \sum_{n=1}^{N} \overline{c}_n (g, g_n) \\ &= \left(g, \sum_{n=1}^{N} c_n g_n \right) = \|g\|^2 \neq 0, \end{aligned}$$

which is absurd. Thus T is well-defined. It is clear that T is linear and that (2) holds.

To show that T is bounded, define a function μ on H by writing

$$\mu(g) = f,$$

where f is the unique solution to (1). We claim that μ is a bounded linear operator. Linearity is easy: if h_1 and h_2 are vectors in H and if a_1 and a_2 are scalars, then for every n

$$\begin{aligned}(\mu(a_1h_1 + a_2h_2), f_n) &= (a_1h_1 + a_2h_2, g_n) \\ &= a_1(h_1, g_n) + a_2(h_2, g_n) \\ &= a_1(\mu(h_1), f_n) + a_2(\mu(h_2), f_n) \\ &= (a_1\mu(h_1) + a_2\mu(h_2), f_n).\end{aligned}$$

Since $\{f_n\}$ is complete, it follows that

$$\mu(a_1h_1 + a_2h_2) = a_1\mu(h_1) + a_2\mu(h_2)$$

and μ is linear. Suppose now that $\{h_n\}$ is a sequence in H such that

$$h_n \to h \quad \text{and} \quad \mu(h_n) \to k.$$

Then

$$(h, g_i) = \lim_{n\to\infty} (h_n, g_i) = \lim_{n\to\infty} (\mu(h_n), f_i) = (k, f_i)$$

for all i, and hence by the definition of μ we have $k = \mu(h)$. Thus μ is bounded by the closed graph theorem.

Let f be an element in Y. Then for every $g \in H$,

$$|(Tf, g)| = |(f, \mu(g))| \leqq \|f\| \cdot \|\mu\| \cdot \|g\|,$$

and hence

$$\|Tf\| \leqq \|f\| \cdot \|\mu\|.$$

Thus T is bounded on Y. Since Y is a dense subspace, we may extend T to be a bounded linear operator on all of H.

Similarly, since (1) can be solved uniquely for g, given f, there exists a bounded linear operator $S: H \to H$ such that

$$Sg_n = f_n \quad (n = 1, 2, 3, \ldots). \tag{3}$$

The proof is over. By combining (2) and (3), we see that $ST = I = TS$, since both $\{f_n\}$ and $\{g_n\}$ are complete. Therefore, T is invertible, and $\{f_n\}$ and $\{g_n\}$ are equivalent bases for H. ∎

The following important theorem is now immediate.

Theorem 8. *A complete sequence of vectors in a separable Hilbert space is a Riesz basis if and only if its moment space is equal to* l^2.

PROBLEMS

1. Deduce Theorem 8 from Theorem 3.

2. Let $\{f_n\}$ and $\{g_n\}$ be two *complete* sequences of elements from a separable Hilbert space H, and suppose that $\sum_{n=1}^{\infty} c_n f_n$ is convergent if and only if $\sum_{n=1}^{\infty} c_n g_n$ is convergent. Does it follow that $\{f_n\}$ and $\{g_n\}$ are equivalent (cf. Theorem 1.7)?

3. Let $\{f_1, f_2, f_3, \ldots\}$ be a complete sequence of nonzero vectors in a separable Hilbert space H. Prove that $\{f_1, f_2, f_3, \ldots\}$ is a basis for H if and only if the moment problem

$$(f, f_n) = c_n \quad (n = 1, 2, 3, \ldots)$$

admits a (unique) solution f for every sequence $\{c_1, c_2, c_3, \ldots\}$ satisfying the following condition: $\sum_{n=1}^{\infty} a_n c_n$ is convergent whenever $\sum_{n=1}^{\infty} a_n f_n$ is convergent.

5 INTERPOLATION IN THE PALEY–WIENER SPACE: FUNCTIONS OF SINE TYPE

Theorem 8 has far-reaching implications for the study of nonharmonic Fourier series in $L^2[-\pi, \pi]$. Thus for a system

$$\{e^{i\lambda_n t}\}$$

of complex exponentials to form a Riesz basis for $L^2[-\pi, \pi]$ it is necessary and sufficient that the trigonometric moment problem

$$\frac{1}{2\pi} \int_{-\pi}^{\pi} \phi(t) e^{-i\lambda_n t}\, dt = c_n \quad (n = 1, 2, 3, \ldots) \tag{1}$$

admit a *unique* solution $\phi \in L^2[-\pi, \pi]$ for each square-summable sequence $\{c_n\}$ of scalars, *and for these sequences only*. By virtue of the isomorphism between $L^2[-\pi, \pi]$ and the Paley–Wiener space, (1) is equivalent to the interpolation problem

$$f(\lambda_n) = c_n \quad (n = 1, 2, 3, \ldots), \tag{2}$$

where $f \in \boldsymbol{P}$. Since $\boldsymbol{P}$ is richly endowed both with a Hilbert space structure and with a distinguished class of elements, problem (2) is often more readily solved than problem (1). In what follows we shall exploit this duality to great advantage.

Definition. *A sequence $\{\lambda_1, \lambda_2, \lambda_3, \ldots\}$ of real or complex numbers is said to be an* ***interpolating sequence*** *for $\boldsymbol{P}$ if the set of all sequences*

$$\{f(\lambda_1), f(\lambda_2), f(\lambda_3), \ldots\},$$

where f ranges over $\boldsymbol{P}$, coincides with l^2. If, in addition, the system $\{e^{i\lambda_n t}\}$ is complete in $L^2[-\pi, \pi]$, then (2) has exactly one solution, provided $\{c_n\} \in l^2$, and in this case we shall call $\{\lambda_1, \lambda_2, \lambda_3, \ldots\}$ a ***complete interpolating sequence****.*

A complete interpolating sequence is clearly "maximal" in the sense that it is not contained in any larger interpolating sequence. The converse is also true (Problem 1).

In the present setting, Theorem 8 takes the following form.

Theorem 9. *A system $\{e^{i\lambda_n t}\}$ of complex exponentials is a Riesz basis for $L^2[-\pi, \pi]$ if and only if $\{\lambda_n\}$ is a complete interpolating sequence for $\boldsymbol{P}$.*

APPLICATION: FUNCTIONS OF SINE TYPE

That the zeros of $\sin \pi z$ appear in the natural basis for $L^2[-\pi, \pi]$ is not entirely fortuitous. In fact, $\sin \pi z$ is but one — the prototype, to be sure — of a large class of entire functions of exponential type whose zeros $\{\lambda_n\}$ give rise to bases $\{e^{i\lambda_n t}\}$ of complex exponentials.

Definition. *An entire function $f(z)$ of exponential type π is said to be of* ***sine type*** *if*

(i) the zeros of $f(z)$ are separated†*, and*
(ii) there exist positive constants A, B, and H such that

$$Ae^{\pi|y|} \leqq |f(x+iy)| \leqq Be^{\pi|y|} \tag{3}$$

whenever x and y are real and $|y| \geqq H$.

According to the definition, a function of sine type is bounded on the real axis and so must have infinitely many zeros. The zeros are all simple and lie in a strip parallel to the real axis.

A large class of examples of entire functions that satisfy part (ii) of the definition is provided by the Fourier–Stieltjes integrals

$$\int_{-\pi}^{\pi} e^{izt}\, d\omega(t),$$

where $\omega(t)$ is of bounded variation on the interval $[-\pi, \pi]$ and has a jump discontinuity at each endpoint (Problem 3). Clearly, $\sin \pi z$ is of this form: simply choose $\omega(t) = 0$ for $-\pi < t < \pi$ and $\omega(\pm\pi) = 1/2i$.

We shall establish the following impressive theorem.

† This requirement is sometimes omitted from the definition.

Theorem 10. *If $\{\lambda_n\}$ is the set of zeros of a function of sine type, then the system $\{e^{i\lambda_n t}\}$ is a Riesz basis for $L^2[-\pi, \pi]$.*

The proof will require several preliminary results. The first is a strengthening of condition (3) inside the domain $|y| < H$.

Lemma 2. *Let $f(z)$ be an entire function of exponential type π and suppose that (3) holds whenever $|y|$ is sufficiently large. For each $\varepsilon > 0$ there corresponds a number $m > 0$ such that*

$$|f(x + iy)| \geqq me^{\pi|y|}$$

outside the circles of radius ε centered at the zeros of $f(z)$.

Proof. Let $\lambda_1, \lambda_2, \lambda_3, \ldots$ be the zeros of $f(z)$. If the lemma is false, then there exist positive constants ε and H and a sequence $z_1, z_2, z_3, \ldots$ of points that lie inside the horizontal strip $|y| < H$, but outside the circles $|z - \lambda_n| = \varepsilon$ $(n = 1, 2, 3, \ldots)$ such that

$$\lim_{n\to\infty} f(z_n) = 0.$$

Write $z_n = x_n + iy_n$, with x_n and y_n real. Since $\sup_n |y_n| < \infty$, we may suppose without loss of generality that $\lim_n y_n$ exists; call it y_0.

Next, define a sequence $f_1(z), f_2(z), f_3(z), \ldots$ of entire functions by writing

$$f_n(z) = f(z + x_n), \quad n = 1, 2, 3, \ldots.$$

Since $f(z)$ is bounded in the horizontal strip $\Omega = \{z: |\operatorname{Im} z| < H + \varepsilon\}$, the sequence $\{f_n(z)\}$ is uniformly bounded there and so forms a normal family in Ω. Accordingly, there is a subsequence that converges uniformly on every compact subset of Ω to a limit function $f_0(z)$. To simplify the notation, we shall continue to denote this subsequence by $\{f_n(z)\}$.

Since

$$f_n(iy_n) \to 0 \quad \text{as} \quad n \to \infty,$$

it follows that

$$f_0(iy_0) = 0.$$

Since $f_0(z)$ does not vanish identically (indeed, $|f_n(iH)| \geqq Ae^{\pi H}$ for $n = 1, 2, 3, \ldots$), Hurwitz's theorem implies that all but a finite number of the f_n's $(n = 1, 2, 3, \ldots)$ must have a zero inside the disk $D = \{z: |z - iy_0| < \varepsilon/2\}$. Say

$$f_n(w_n) = 0 \quad \text{with} \quad w_n \in D \quad \text{and} \quad n > n_0.$$

If we set

$$\gamma_n = w_n + x_n \quad \text{for} \quad n > n_0,$$

then $f(\gamma_n) = 0$, so that each γ_n must be one of the λ's. But

$$|z_n - \gamma_n| = |w_n - iy_n| \leqq |w_n - iy_0| + |iy_0 - iy_n| < \frac{\varepsilon}{2} + \frac{\varepsilon}{2} = \varepsilon$$

as soon as n is sufficiently large, and this contradicts the choice of z_n. The contradiction proves the lemma. ■

Corollary 1. *If $f(z)$ is a function of sine type and $\lambda_1, \lambda_2, \lambda_3, \ldots$ are its zeros, then*

$$\inf_n |f'(\lambda_n)| > 0.$$

Proof. Choose $\varepsilon > 0$ so small that

$$|\lambda_n - \lambda_m| > 2\varepsilon \quad \text{whenever} \quad n \neq m.$$

Then the disks $D_n\colon |z - \lambda_n| \leqq \varepsilon$ $(n = 1, 2, 3, \ldots)$ are nonoverlapping, and each function $f(z)/(z - \lambda_n)$ is analytic and free of zeros inside D_n. If $m > 0$ is chosen in accordance with Lemma 2, then

$$\log\left|\frac{f(z)}{z - \lambda_n}\right| \geqq \log\left(\frac{m}{\varepsilon}\right) \quad \text{for} \quad |z - \lambda_n| = \varepsilon \quad (n = 1, 2, 3, \ldots),$$

and the result follows from the minimum principle for harmonic functions. ■

Corollary 2. *If $\{\lambda_n\}$ is the set of zeros of a function of sine type, then the system $\{e^{i\lambda_n t}\}$ is complete in $L^2[-\pi, \pi]$.*

Proof. Let $f(z)$ be a function of sine type with zeros $\lambda_1, \lambda_2, \lambda_3, \ldots$. It is sufficient to show that $\{\lambda_n\}$ is a set of uniqueness for $\boldsymbol{P}$. Suppose therefore that $g(z) \in \boldsymbol{P}$ and $g(\lambda_n) = 0$ for $n = 1, 2, 3, \ldots$. Then the function

$$\phi(z) = \frac{g(z)}{f(z)}$$

is entire. By the corollary to Theorem 2.18,

$$g(z)e^{-\pi|y|} \to 0 \quad \text{as} \quad |z| \to \infty,$$

and hence by Lemma 2, $\phi(z) \to 0$ as $|z| \to \infty$, outside a system of (nonoverlapping) circles $|z - \lambda_n| = \varepsilon$ $(n = 1, 2, 3, \ldots)$. The maximum principle guarantees that $\phi(z) \to 0$ inside the excluded circles as well, and therefore $\phi(z)$ must vanish identically. Hence $g(z)$ vanishes identically, and $\{\lambda_n\}$ is a set of uniqueness for $\boldsymbol{P}$. ■

Let D be the upper half-plane, $\operatorname{Im} z > 0$. We shall make use of the space $H^2(D)$ consisting of all complex-valued functions $f(z)$ that are analytic throughout D and such that

$$\|f\| = \sup_{y>0}\left(\int_{-\infty}^{\infty} |f(x + iy)|^2\, dx\right)^{1/2} < \infty.$$

The following facts about $H^2(D)$ are well known (see Duren [1970, Chap. 11] and Problem 6).

1. If $f \in H^2(D)$, then the *boundary function*

$$f(x) = \lim_{y \to 0} f(x + iy)$$

exists almost everywhere, and

$$\|f\|^2 = \int_{-\infty}^{\infty} |f(x)|^2 \, dx.$$

It follows that an inner product on $H^2(D)$ is defined by setting

$$(f, g) = \int_{-\infty}^{\infty} f(x)\overline{g(x)} \, dx.$$

Endowed with this inner product, $H^2(D)$ is a separable Hilbert space.

2. Every function f in $H^2(D)$ can be recovered from its boundary values by means of a Cauchy integral:

$$f(z) = \frac{1}{2\pi i} \int_{-\infty}^{\infty} \frac{f(t)}{t - z} \, dt, \quad \operatorname{Im} z > 0.$$

3. If $\{\lambda_1, \lambda_2, \lambda_3, \ldots\}$ is a separated sequence of points lying in the horizontal strip

$$0 < m \leqq \operatorname{Im} z \leqq M < \infty,$$

then

$$\sqrt{\sum_{n=1}^{\infty} |f(\lambda_n)|^2} \leqq B\|f\|$$

for some constant B and every element $f \in H^2(D)$.

We are now ready to prove Theorem 10.

Proof of Theorem 10. Let $g(z)$ be a function of sine type with zeros $\lambda_1, \lambda_2, \lambda_3, \ldots$. By Theorem 9 it suffices to show that $\{\lambda_n\}$ is a complete interpolating sequence for the Paley–Wiener space. This means that

(i) the interpolation problem

$$f(\lambda_n) = c_n \quad (n = 1, 2, 3, \ldots)$$

has a solution $f \in \boldsymbol{P}$ whenever $\sum |c_n|^2 < \infty$,

(ii) $\sum_{n=1}^{\infty} |f(\lambda_n)|^2 < \infty$ whenever $f \in \boldsymbol{P}$, and

(iii) the system $\{e^{i\lambda_n t}\}$ is complete in $L^2[-\pi, \pi]$.

Of these three statements, only the first requires proof: the second is valid because $\{\lambda_n\}$ is separated and $\sup_n |\operatorname{Im}\lambda_n| < \infty$, while the third follows from Corollary 2.

Let $\{c_n\}$ belong to l^2. The most natural choice for a function that assumes the value c_n at the point λ_n is the Lagrange interpolation series

$$f(z) = \sum_{n=1}^{\infty} \frac{c_n g(z)}{g'(\lambda_n)(z-\lambda_n)}.$$

We are going to show that the series converges in the norm of $\boldsymbol{P}$, and therefore uniformly in every horizontal strip; accordingly, $f \in \boldsymbol{P}$ and $f(\lambda_n) = c_n$ $(n = 1, 2, 3, \ldots)$.

Let m and n be arbitrary but fixed positive integers, with $m < n$, and write

$$G(z) = \sum_{k=m}^{n} \frac{c_k g(z)}{g'(\lambda_k)(z-\lambda_k)}.$$

Then $G(z) \in \boldsymbol{P}$, and we have only to show that $\|G\| \to 0$ as $m, n \to \infty$.

Begin by choosing positive constants A, B, and H such that

$$Ae^{\pi|y|} \leqq |g(x+iy)| \leqq Be^{\pi|y|}$$

for all values of x and $|y| \geqq H$, and put

$$h(z) = \frac{G(z+2iH)}{g(z+2iH)}.$$

Straightforward calculations show that $h(z) \in H^2(D)$, and hence

$$\begin{aligned}
\|G\| &= \left(\int_{-\infty}^{\infty} |G(x)|^2\, dx\right)^{1/2} \\
&\leqq e^{2\pi H}\left(\int_{-\infty}^{\infty} |G(x+2iH)|^2\right)^{1/2} && \text{(by Theorem 2.16)} \\
&\leqq Be^{4\pi H}\left(\int_{-\infty}^{\infty} |h(x)|^2\right)^{1/2} \\
&= Be^{4\pi H}\|h\| && \text{(by property 1).}
\end{aligned}$$

We estimate the norm of h as follows.

Choose φ in the unit sphere of $H^2(D)$ such that $\|h\| = (h, \varphi)$. Then

$$\begin{aligned}
\|h\| &= \int_{-\infty}^{\infty} h(x)\overline{\varphi(x)}\, dx \\
&= -2\pi i \sum_{m}^{n} \frac{c_k \overline{\varphi}(\overline{\lambda}_k + 2iH)}{g'(\lambda_k)} && \text{(by property 2)} \\
&\leqq 2\pi \sqrt{\sum_{m}^{n} |c_k/g'(\lambda_k)|^2}\sqrt{\sum_{m}^{n} |\varphi(\overline{\lambda}_k + 2iH)|^2}.
\end{aligned}$$

By Corollary 1 we may choose $\varepsilon > 0$ so small that $|g'(\lambda_k)| \geqq \varepsilon$ for $k = 1, 2, 3, \ldots$, and by property 3 we may choose B so large that the second sum on the right is at most B^2. It follows that

$$\|h\| \leqq \frac{2\pi B}{\varepsilon}\sqrt{\sum_m^n |c_k|^2},$$

and hence

$$\|G\| \leqq C\sqrt{\sum_m^n |c_k|^2},$$

where C is independent of m and n. Since $\{c_n\} \in l^2$, the right side approaches zero as $m, n \to \infty$. Thus $\|G\| \to 0$. ∎

PROBLEMS

1. Prove that a *maximal* interpolating sequence for $\boldsymbol{P}$ is complete.
2. Let $\{\lambda_n\}$ be an interpolating sequence for $\boldsymbol{P}$ and $f_n(z)$ the unique element of $\boldsymbol{P}$ of minimal norm for which $f_n(\lambda_m) = \delta_{mn}$. Show that if $\{c_n\} \in l^2$, then the series
$$\sum_{n=1}^{\infty} c_n f_n(z)$$
converges in $\boldsymbol{P}$ to a function $f(z)$ for which $f(\lambda_n) = c_n$ for every n.
3. **(Sedleckiĭ)** Let
$$f(z) = \int_{-\pi}^{\pi} e^{izt}\, d\omega(t),$$
where $\omega(t)$ is of bounded variation on $[-\pi, \pi]$ and has a jump at $t = \pi$.
(a) Prove that there exist positive constants A and H such that
$$|f(x+iy)| \geqq Ae^{\pi|y|} \tag{4}$$
for all values of x and $y < -H$. (*Hint*: Let h be the value of the jump of $\omega(t)$ at $t = \pi$. Given $\varepsilon > 0$, choose $\delta > 0$ so that the total variation of $\omega(t)$ on the interval $[\pi - \delta, \pi)$ does not exceed ε. Then
$$f(z) = he^{i\pi z} + \left(\int_{-\pi}^{0} + \int_{0}^{\pi-\delta} + \int_{\pi-\delta}^{\pi-}\right) e^{izt}\, d\omega(t).$$
If $y < 0$, then
$$|f(x+iy)| \geqq e^{\pi|y|}(|h| - Ve^{-\pi|y|} - Ve^{-\delta|y|} - \varepsilon),$$
where V is the total variation of $\omega(t)$ on $[-\pi, \pi]$.)

(b) Prove that if (4) holds for a sequence of points iy_n of the imaginary axis such that $y_n \to -\infty$, then $\omega(t)$ has a jump at $t = \pi$.

4. Prove or disprove: Two functions of sine type with the same set of zeros can differ at most by a multiplicative constant.

5. Show that if $f(z) \in \boldsymbol{P}$, then $f(z)e^{i\pi z} \in H^2(D)$.

6. Establish property 3 for functions belonging to $H^2(D)$. (*Hint*: Choose $\varepsilon > 0$ so small that the disks $D_n\colon |z - z_n| \leqq \varepsilon$ are disjoint and lie in the upper half-plane, $\operatorname{Im} z > 0$. Then

$$\sum_n \iint_{D_n} |f(z)|^2\, dx\, dy \leqq \int_0^M \int_{-\infty}^{\infty} |f(x+iy)|^2\, dx\, dy \leqq M\|f\|^2.)$$

7. **(Interpolation in $\boldsymbol{E}_\pi^p$)** Let $\{\lambda_n\}$ be the zeros of a function $g(z)$ of sine type, $1 < p < \infty$, and $\{c_n\}$ a sequence in l^p.

(a) Show that the series

$$\sum_{n=1}^{\infty} \frac{c_n g(z)}{g'(\lambda_n)(z - \lambda_n)}$$

converges in $\boldsymbol{E}_\pi^p$ to a function $f(z)$ for which $f(\lambda_n) = c_n$ for all n. (For the definition of $\boldsymbol{E}_\pi^p$, see Problem 5 of Section 2.3, Part Two.)

(b) Take $g(z) = \sin \pi z$. Exhibit sequences in $l^p (p = 1, \infty)$ for which the equations

$$f(n) = c_n \quad (n = 0, \pm 1, \pm 2, \ldots)$$

have no corresponding solution in $\boldsymbol{E}_\pi^p$. (*Hint*: See Problem 8, Section 2. By definition, $\boldsymbol{E}_\pi^\infty = \boldsymbol{B}_\pi$.)

6 INTERPOLATION IN THE PALEY–WIENER SPACE: STABILITY

By definition, a sequence $\{\lambda_1, \lambda_2, \lambda_3, \ldots\}$ of scalars is an interpolating sequence for the Paley–Wiener space provided that (i) the equations $f(\lambda_n) = c_n$ $(n = 1, 2, 3, \ldots)$ have at least one solution $f \in \boldsymbol{P}$ for each sequence $\{c_n\}$ in l^2, and (ii) $\sum |f(\lambda_n)|^2 < \infty$ for every $f \in \boldsymbol{P}$. The first condition is of course the more difficult one to verify; given the first, the second frequently comes gratis.

Proposition 3. *Let $\lambda_1, \lambda_2, \lambda_3, \ldots$ lie in a strip parallel to the real axis. If the equations*

$$f(\lambda_n) = c_n \quad (n = 1, 2, 3, \ldots)$$

admit at least one solution $f \in \boldsymbol{P}$ for each square-summable sequence $\{c_n\}$ of scalars, then $\{\lambda_n\}$ is separated, and so

$$\sum |f(\lambda_n)|^2 < \infty \quad \textit{for every } f \in \boldsymbol{P}.$$

Proof. Since $\boldsymbol{P}$ is a functional Hilbert space, there is a sequence $K_1, K_2, K_3, \ldots$ of elements of $\boldsymbol{P}$ such that

$$f(\lambda_n) = (f, K_n)$$

for $n = 1, 2, 3, \ldots$ and every $f \in \boldsymbol{P}$. Indeed, if K is the reproducing kernel of $\boldsymbol{P}$, then

$$K_n(z) = K(z, \lambda_n) = \frac{\sin \pi(z - \overline{\lambda}_n)}{\pi(z - \overline{\lambda}_n)}. \tag{1}$$

By assumption, $\{K_n\}$ is a *Riesz–Fischer sequence* in $\boldsymbol{P}$, and hence the unit ball of l^2 can be interpolated in a uniformly bounded way. In particular, there exist a constant M and a sequence $f_1, f_2, f_3, \ldots$ of elements of $\boldsymbol{P}$ such that

$$f_n(\lambda_m) = \delta_{mn} \quad \text{and} \quad \|f_n\| \leqq M.$$

Choose H so large that $|\operatorname{Im} \lambda_n| \leqq H$ for every n. If $m \neq n$, then

$$1 = f_n(\lambda_n) - f_n(\lambda_m) = \int_{\lambda_m}^{\lambda_n} f_n'(z)\, dz,$$

and so

$$1 \leqq |\lambda_n - \lambda_m| \cdot \sup |f_n'(z)|,$$

where the supremum is taken over all points z on the line segment joining λ_m and λ_n. For these points, $|\operatorname{Im} z| \leqq H$, and it follows from known properties of $\boldsymbol{P}$ that

$$|f_n'(z)| \leqq e^{\pi H} \|f_n'\| \leqq \pi e^{\pi H} \|f_n\| \leqq M\pi e^{\pi H}.$$

Thus

$$1 \leqq M\pi e^{\pi H} |\lambda_n - \lambda_m| \quad \text{whenever} \quad n \neq m,$$

and $\{\lambda_n\}$ is separated. Since $\sup_n |\operatorname{Im} \lambda_n| < \infty$, the *Remark* following Theorem 2.17 shows that $\sum |f(\lambda_n)|^2 < \infty$ whenever $f \in \boldsymbol{P}$. ∎

Corollary 1. *The points of an interpolating sequence lie in a strip parallel to the real axis and are separated.*

Proof. Let $\{\lambda_n\}$ be an interpolating sequence for $\boldsymbol{P}$, and let $\{K_n\}$ be the corresponding sequence of kernel functions (1). By assumption, $\{K_n\}$ is a *Bessel sequence* in $\boldsymbol{P}$, and so by Proposition 2

$$\sum |f(\lambda_n)|^2 \leqq B\|f\|^2$$

for some constant B and every function $f \in \boldsymbol{P}$. It follows readily from this that

$$\left\|\sum c_n e^{i\lambda_n t}\right\|^2 \leqq B \sum |c_n|^2$$

whenever $\{c_n\} \in l^2$ (first observe that $\left\|\sum c_n K_n\right\|^2 \leqq B \sum |c_n|^2$, and then use the Fourier isometry). We conclude that

$$\left\|e^{i\lambda_n t}\right\| \leqq \sqrt{B} \quad \text{for every } n,$$

and hence that

$$\sup_n |\operatorname{Im} \lambda_n| < \infty.$$

This proves the first assertion; the second follows from Proposition 3. ■

Corollary 2. *If the system of exponentials $\{e^{i\lambda_n t}\}$ is a Riesz basis for $L^2[-\pi, \pi]$, then the points λ_n lie in a strip parallel to the real axis and are separated.*

The goal of the present section is to show that interpolating sequences in the Paley–Wiener space are *stable* under sufficiently small perturbations of their elements. It is the first step towards a general stability criterion for Riesz bases of complex exponentials (see Section 8). We begin by proving two lemmas, which are of interest in their own right.

Lemma 3. *Let $\{\lambda_1, \lambda_2, \lambda_3, \ldots\}$ be a sequence of scalars such that*

$$\sum |f(\lambda_n)|^2 \leqq B\|f\|^2 \quad \textit{for every } f \in \boldsymbol{P}.$$

If $\{\mu_1, \mu_2, \mu_3, \ldots\}$ satisfies

$$|\lambda_n - \mu_n| \leqq L, \quad n = 1, 2, 3, \ldots,$$

then for every f,

$$\sum |f(\lambda_n) - f(\mu_n)|^2 \leqq B(e^{\pi L} - 1)^2 \|f\|^2.$$

Proof. Let f be an element of $\boldsymbol{P}$. By expanding f in a Taylor series about λ_n, we find that

$$f(\mu_n) - f(\lambda_n) = \sum_{k=1}^{\infty} \frac{f^{(k)}(\lambda_n)}{k!} (\mu_n - \lambda_n)^k \quad (n = 1, 2, 3, \ldots).$$

If ρ is an arbitrary positive number, then by multiplying and dividing the summand by ρ^k we find also that

$$|f(\mu_n) - f(\lambda_n)|^2 \leqq \sum_{k=1}^{\infty} \frac{|f^{(k)}(\lambda_n)|^2}{\rho^{2k} k!} \sum_{k=1}^{\infty} \frac{\rho^{2k} |\mu_n - \lambda_n|^{2k}}{k!}. \tag{2}$$

Since $\boldsymbol{P}$ is closed under differentiation and $\|f'\| \leqq \pi\|f\|$, it follows that

$$\sum_n |f^{(k)}(\lambda_n)|^2 \leqq B\|f^{(k)}\|^2 \leqq B\pi^{2k}\|f\|^2 \quad (k = 1, 2, 3, \ldots). \tag{3}$$

By combining (2) and (3), we obtain

$$\sum |f(\lambda_n) - f(\mu_n)|^2 \leqq B\|f\|^2 \sum_{k=1}^{\infty} \frac{\pi^{2k}}{\rho^{2k}k!} \sum_{k=1}^{\infty} \frac{(\rho L)^{2k}}{k!}$$

$$= B\|f\|^2 (e^{\pi^2/\rho^2} - 1)(e^{\rho^2 L^2} - 1)$$

since $|\lambda_n - \mu_n| \leqq L$. The result follows by choosing $\rho = \sqrt{\pi/L}$. ∎

Remark. It should be noted that the lemma applies whenever $\{\lambda_n\}$ is separated and $\{\operatorname{Im} \lambda_n\}$ is bounded, and hence whenever $\{\lambda_n\}$ is an interpolating sequence.

The next lemma provides a convenient criterion for insuring that a bounded linear transformation maps one Banach space *onto* another.

Lemma 4. *Let T be a bounded linear transformation of the Banach space X into the Banach space Y. Suppose that there exist positive constants M and ε, $0 < \varepsilon < 1$, with the following property: for each vector y in the unit ball of Y there is a vector $x \in X$ such that*

$$\|x\| \leqq M \quad \textit{and} \quad \|Tx - y\| \leqq \varepsilon.$$

Then T maps X **onto** *Y.*

Proof. First observe that if $y \in Y$ and $\|y\| \leqq \lambda$, then there is a vector $x \in X$ such that $\|x\| \leqq M\lambda$ and $\|Tx - y\| \leqq \varepsilon\lambda$. Let y be an arbitrary element in the unit ball of Y. It is clearly sufficient to show that $y = Tx$ for some x.

Begin by choosing x_1 such that $\|x_1\| \leqq M$ and $\|y - Tx_1\| \leqq \varepsilon$. Next, with $y - Tx_1$ in place of y and ε in place of λ, choose x_2 such that $\|x_2\| \leqq M\varepsilon$ and $\|(y - Tx_1) - Tx_2\| \leqq \varepsilon^2$. Proceed by induction. The result is a sequence $\{x_1, x_2, x_3, \ldots\}$ of vectors of X such that

$$\|x_n\| \leqq M\varepsilon^{n-1} \quad \text{and} \quad \left\| y - \sum_{i=1}^{n} Tx_i \right\| \leqq \varepsilon^n. \tag{4}$$

Since $0 < \varepsilon < 1$, we have $\sum \|x_n\| \leqq \sum M\varepsilon^{n-1} = M/(1 - \varepsilon)$. Thus the series $\sum x_n$ is absolutely convergent, and so there is an element x in X to which the partial sums of the series converge. Since $\varepsilon^n \to 0$, (4) shows that $y = Tx$. ∎

We are now in a position to prove that interpolating sequences in the Paley–Wiener space are stable.

Theorem 11. *Let $\{\lambda_1, \lambda_2, \lambda_3, \ldots\}$ be an interpolating sequence for $\boldsymbol{P}$. There exists a positive constant L with the property that $\{\mu_1, \mu_2, \mu_3, \ldots\}$ is also an interpolating sequence for $\boldsymbol{P}$ whenever $|\lambda_n - \mu_n| \leqq L$ for every n.*

Proof. Let $\mu_1, \mu_2, \mu_3, \ldots$ be complex scalars such that $|\lambda_n - \mu_n| \leqq L$ for every n; the value of L will be specified later. Let T be the induced linear transformation of $\boldsymbol{P}$ into the vector space of all sequences defined by

$$Tf = \{f(\mu_1), f(\mu_2), f(\mu_3), \ldots\}.$$

The assertion is that if L is sufficiently small then $T\boldsymbol{P} = l^2$.

Since $\{\lambda_n\}$ is an interpolating sequence, $\sum |f(\lambda_n)|^2 \leqq B\|f\|^2$ for some constant B and every $f \in \boldsymbol{P}$, and hence the inequality

$$\sum |f(\lambda_n) - f(\mu_n)|^2 \leqq B(e^{\pi L} - 1)^2 \|f\|^2 \tag{5}$$

of Lemma 3 is valid. This shows that T is a bounded linear transformation from $\boldsymbol{P}$ *into* l^2. By Proposition 2 there is a constant M such that the system of equations

$$f(\lambda_n) = c_n \quad (n = 1, 2, 3, \ldots) \tag{6}$$

has at least one solution $f \in \boldsymbol{P}$ satisfying

$$\|f\|^2 \leqq M \sum |c_n|^2, \tag{7}$$

provided that $\{c_n\} \in l^2$.

The proof of the theorem is at hand. Let $c = \{c_n\}$ be an arbitrary element in the unit ball of l^2. If f is chosen in accordance with (6) and (7), then (5) becomes

$$\|Tf - c\|^2 \leqq B(e^{\pi L} - 1)^2 M.$$

By choosing L sufficiently small, we can make the right side less than 1, and the assertion $T\boldsymbol{P} = l^2$ follows from Lemma 4. ■

PROBLEMS

1. Let $\{\lambda_1, \lambda_2, \lambda_3, \ldots\}$ be a sequence of points that lie in a strip parallel to the real axis and let T be the induced linear transformation from $\boldsymbol{P}$ into the vector space of all sequences defined by

$$Tf = \{f(\lambda_1), f(\lambda_2), f(\lambda_3), \ldots\}.$$

Show that if $T\boldsymbol{P} \supset l^1$, then $\{\lambda_n\}$ is separated, and hence $T\boldsymbol{P} \subset l^2$.

2. Exhibit a bounded linear operator T on l^2 with the following two properties:

(1) There is a constant M such that whenever $y \in l^2$ and $\|y\| \leqq 1$, we can find an element $x \in l^2$ with $\|x\| \leqq M$ and $\|Tx - y\| < 1$.

(2) T is *not* onto.

3. Let $\{\lambda_n\}$ be a sequence of points that lie in a strip parallel to the real axis and suppose that

$$\sum |f(\lambda_n)|^2 < \infty \quad \text{for every } f \in \boldsymbol{P}.$$

Prove that it is possible to partition $\{\lambda_n\}$ into a finite number of separated sequences.

4. Let $\{\lambda_n\}$ be a complete interpolating sequence for $\boldsymbol{P}$ and suppose that $\operatorname{Re}\lambda_n \leq \operatorname{Re}\lambda_{n+1}$ for every n. Prove that

$$\sup_n |\lambda_{n+1} - \lambda_n| < \infty.$$

7 THE THEORY OF FRAMES

We have seen that the study of Riesz bases in a separable Hilbert space H can be reduced to the study of an abstract moment problem:

$$(f, f_n) = c_n \quad (n = 1, 2, 3, \ldots).$$

In this section we shall introduce the notion of a "frame", which will further elucidate the problem of moments and at the same time will provide yet another characterization of Riesz bases.

Definition. *A sequence of distinct vectors $\{f_1, f_2, f_3, \ldots\}$ belonging to a separable Hilbert space H is said to be a **frame** if there exist positive constants A and B such that*

$$A\|f\|^2 \leqq \sum_{n=1}^{\infty} |(f, f_n)|^2 \leqq B\|f\|^2 \tag{1}$$

*for every $f \in H$. The numbers A and B are called **bounds** for the frame.*

It is clear that a frame is a complete set of vectors since the relations $(f, f_n) = 0$, $n = 1, 2, 3, \ldots$, imply that $f = 0$. Notice also that the second inequality in (1) is just the condition that $\{f_n\}$ be a Bessel sequence with bound B. In practice, it is the first inequality that is difficult to verify.

Example 1. If $\{f_n\}$ is a complete orthonormal sequence in H, then Parseval's identity shows that (1) holds with $A = B = 1$.

Example 2 (Nonharmonic Fourier Frames). A system of complex exponentials $\{e^{i\lambda_n t}\}$ is a frame in $L^2[-\pi, \pi]$ provided that

$$A\int_{-\pi}^{\pi} |\phi(t)|^2\,dt \leqq \sum_n \left|\int_{-\pi}^{\pi} \phi(t)e^{-i\lambda_n t}\,dt\right|^2 \leqq B\int_{-\pi}^{\pi} |\phi(t)|^2\,dt$$

for every $\phi \in L^2[-\pi, \pi]$. By virtue of the Paley–Wiener theorem, this is equivalent to

$$A \int_{-\infty}^{\infty} |f(x)|^2\, dx \leqq \sum_n |f(\lambda_n)|^2 \leqq B \int_{-\infty}^{\infty} |f(x)|^2\, dx$$

for every function $f(z)$ belonging to the Paley–Wiener space.

We associate with a given frame $\{f_1, f_2, f_3, \ldots\}$ a bounded linear operator T on H defined by

$$Tf = \sum_{n=1}^{\infty} (f, f_n) f_n \tag{2}$$

(T is a bounded operator because $\{f_n\}$ is a Bessel sequence: $\|Tf\|^2 \leqq B \sum |(f, f_n)|^2 \leqq B^2 \|f\|^2$).

Assertion: T is invertible. Since

$$(Tf, f) = \sum_{n=1}^{\infty} |(f, f_n)|^2,$$

it follows that

$$A\|f\|^2 \leqq (Tf, f) \leqq \|Tf\| \cdot \|f\|,$$

and so

$$A\|f\| \leqq \|Tf\|.$$

This shows that T is bounded from below, and so must be one-to-one. But T is self-adjoint (indeed, positive), and an operator whose adjoint is bounded from below must be onto (see Taylor [1958, p. 234]). Thus T is invertible.

Throughout this section T will denote the transformation defined by (2).

Definition. *A frame that ceases to be a frame when any one of its elements is removed is said to be an* ***exact frame****.*

We are going to show that the class of Riesz bases and the class of exact frames are identical. It is convenient to first establish the following two lemmas.

Lemma 5. *Let $\{f_1, f_2, f_3, \ldots\}$ be a frame in a separable Hilbert space H and f an arbitrary element of H. Then there is a unique moment sequence $\{a_1, a_2, a_3, \ldots\}$ such that*

$$f = \sum_{n=1}^{\infty} a_n f_n,$$

and

$$a_n = (g, f_n), \quad \textit{where} \quad Tg = f.$$

If $\{b_1, b_2, b_3, \ldots\}$ *is any other sequence such that* $f = \sum_{n=1}^{\infty} b_n f_n$, *then*

$$\sum_{n=1}^{\infty} |b_n|^2 = \sum_{n=1}^{\infty} |a_n|^2 + \sum_{n=1}^{\infty} |a_n - b_n|^2. \tag{3}$$

Proof. The first statement is immediate since T is invertible and $f = T(T^{-1}f)$. Suppose now that

$$\sum_{n=1}^{\infty} a_n f_n = \sum_{n=1}^{\infty} b_n f_n.$$

Taking the inner product of both sides with g, we find

$$\sum_{n=1}^{\infty} |a_n|^2 = \sum_{n=1}^{\infty} \overline{a}_n b_n,$$

and (3) follows readily. ■

Lemma 6. *The removal of a vector from a frame leaves either a frame or an incomplete set.*

Proof. Let $\{f_n\}$ be a frame in H, with bounds A and B, and suppose that f_m is removed. By Lemma 5

$$f_m = \sum_{n=1}^{\infty} a_n f_n,$$

where $a_n = (g_m, f_n)$ and $Tg_m = f_m$. We consider the two cases $a_m = 1$ and $a_m \neq 1$, showing that in the first case the system $\{f_n\}_{n \neq m}$ is incomplete, while in the second case it remains a frame.

If $a_m = 1$, then

$$\sum_{n \neq m} a_n f_n = 0,$$

and so

$$\sum_{n=1}^{\infty} a'_n f_n = 0,$$

where $a'_n = a_n$ when $n \neq m$ and $a'_m = 0$. Applying Lemma 5, with every b_n equal to zero, we find

$$\sum_{n=1}^{\infty} |a'_n|^2 = 0,$$

and hence $a_n = 0$ for $n \neq m$. Thus g_m is orthogonal to f_n for $n \neq m$. Since $(g_m, f_m) = a_m = 1$, it follows that $g_m \neq 0$ and hence that $\{f_n\}_{n \neq m}$ is incomplete.

Suppose now that $a_m \neq 1$. Then

$$f_m = \frac{1}{1-a_m} \sum_{n \neq m} a_n f_n.$$

Consequently, for all f in H

$$|(f, f_m)|^2 \leqq \frac{1}{|1-a_m|^2} \sum_{n \neq m} |a_n|^2 \sum_{n \neq m} |(f, f_n)|^2,$$

and so

$$\sum_{n=1}^{\infty} |(f, f_n)|^2 \leqq C \sum_{n \neq m} |(f, f_n)|^2,$$

where $C = 1 + |1-a_m|^{-2} \sum_{n \neq m} |a_n|^2$. Since $\{f_n\}$ is a frame, we conclude that

$$A_1 \|f\|^2 \leqq \sum_{n \neq m} |(f, f_n)|^2 \leqq B_1 \|f\|^2,$$

where $A_1 = A/C$ and $B_1 = B$. Thus the system $\{f_n\}_{n \neq m}$ is a frame. ∎

Remark. An examination of the proof reveals that if $\{f_n\}$ is an exact frame and $g_n = T^{-1} f_n$, then $\{f_n\}$ and $\{g_n\}$ are biorthogonal.

We are now in a position to give the following characterization of Riesz bases.

Theorem 12. *A sequence of vectors belonging to a separable Hilbert space H is a Riesz basis if and only if it is an exact frame.*

Proof. Let $\{f_n\}$ be a Riesz basis for H. Then it is complete and has l^2 as its moment space. Accordingly, the mapping

$$f \rightarrow \{(f, f_n)\}$$

defines a one-to-one linear transformation of H onto l^2. Since it is bounded (by Proposition 2), its inverse is also bounded (by the open mapping theorem), and the frame condition follows at once. Since the removal of a vector from a basis leaves an incomplete set, $\{f_n\}$ is an exact frame.

Suppose now that $\{f_n\}$ is an exact frame in H, with bounds A and B, and let f be an arbitrary element of H. Then by Lemma 5

$$f = \sum_{n=1}^{\infty} a_n f_n,$$

where $a_n = (g, f_n)$ and $g = T^{-1}f$. To prove that this representation is unique, suppose that

$$f = \sum_{n=1}^{\infty} b_n f_n.$$

If we define g_n by setting $g_n = T^{-1}f_n$, then $\{f_n\}$ and $\{g_n\}$ are biorthogonal (by the *Remark* following Lemma 6), and hence

$$a_n = (g, f_n) = (g, Tg_n) = (Tg, g_n) = (f, g_n) = b_n$$

since T is self-adjoint. Thus $\{f_n\}$ is a basis for H.

It remains only to show that $\{f_n\}$ is a Riesz basis. By Theorem 1.7 this will follow if we can show that the series $\sum c_n f_n$ is convergent if and only if $\sum |c_n|^2 < \infty$.

Let the series $\sum c_n f_n$ be convergent, say to an element f of H. Then $c_n = (g, f_n)$, where $g = T^{-1}f$, and so by the frame condition

$$\sum |c_n|^2 = \sum |(g, f_n)|^2 \leqq B\|g\|^2 < \infty.$$

Suppose now that $\{c_n\}$ is any sequence of scalars such that $\sum |c_n|^2 < \infty$. Then for $n \geqq m$ we have by Theorem 3

$$\left\|\sum_m^n c_i f_i\right\|^2 \leqq B \sum_m^n |c_i|^2$$

since $\{f_n\}$ is a Bessel sequence with bound B. Since the right side tends to zero as $m, n \to \infty$, the partial sums of the series $\sum c_n f_n$ form a Cauchy sequence, and hence the series is convergent. ∎

The literature on the theory and applications of frames and Riesz bases is now extensive, due in part to the "wavelets revolution" of the past 15 years. For additional results and references, see the *Problems* at the end of the section as well as the *Notes*.

PROBLEMS

1. Let $\{f_1, f_2, f_3, \ldots\}$ be a frame in a separable Hilbert space H and put $g_n = T^{-1}f_n$. Prove that $\{g_1, g_2, g_3, \ldots\}$ is also a frame and that for every f,

$$f = \sum_{n=1}^{\infty} (f, g_n) f_n = \sum_{n=1}^{\infty} (f, f_n) g_n.$$

2. Prove that if $\{f_1, f_2, f_3, \ldots\}$ is an *exact frame*, with bounds A and B, then

$$A \sum |c_n|^2 \leqq \left\|\sum c_n f_n\right\|^2 \leqq B \sum |c_n|^2$$

whenever $\{c_n\} \in l^2$.

3. **(Duffin–Schaeffer)** Let $\{f_1, f_2, f_3, \ldots\}$ be a frame, with bounds A and B, and let $\rho = 2/(A+B)$. If h is an arbitrary vector, we define a sequence $h^{(0)}, h^{(1)}, h^{(2)}, \ldots$ recursively by setting $h^{(0)} = h$ and

$$h^{(k+1)} = h^{(k)} - \rho \sum_{n=1}^{\infty} (h^{(k)}, f_n) f_n \quad \text{for} \quad k = 0, 1, 2, \ldots.$$

Let

$$\beta_n^{(k)} = \rho(h^{(0)} + h^{(1)} + \cdots + h^{(k-1)}, f_n)$$

and

$$h_k = \sum_{n=1}^{\infty} \beta_n^{(k)} f_n.$$

Prove that

$$\|h - h_k\| \leqq \left(\frac{B-A}{B+A}\right)^k \|h\|.$$

(*Hint*: If $S = I - \rho T$, then

$$|(Sh, h)| \leqq \theta \|h\|^2, \quad \theta = \frac{B-A}{B+A}.$$

Conclude that $\|h^{(r)}\| \leqq \theta^r \|h\|$. Now add the relations

$$h^{(r+1)} - h^{(r)} = -\rho T(h^{(r)})$$

for $r = 0, 1, \ldots, k-1$.)

4. Let $\{f_1, f_2, f_3, \ldots\}$ be a frame in a separable Hilbert space H and suppose that $\{c_1, c_2, c_3, \ldots\}$ is an arbitrary square-summable sequence of scalars. Prove that there is a unique element f in H minimizing

$$\sum_{n=1}^{\infty} |c_n - (f, f_n)|^2.$$

5. Prove that every finite subset of a Hilbert space is a frame in its linear span.

6. A frame $\{f_1, f_2, f_3, \ldots\}$, with bounds A and B, is said to be *tight* if $A = B$. A tight frame is said to be *normalized* if $A = B = 1$.

(a) Prove that for a tight frame, the representation

$$f = \frac{1}{A} \sum (f, f_n) f_n$$

is valid in norm for every f.

(b) Is every normalized tight frame an orthonormal basis?

8 THE STABILITY OF NONHARMONIC FOURIER SERIES

In this section we shall discuss the stability of the class of Riesz bases $\{e^{i\lambda_n t}\}$ in $L^2[-\pi, \pi]$, first under "small" displacements of the λ_n's and then under more general "vertical" displacements. As a result, we shall find that Kadec's $\frac{1}{4}$-theorem can be dramatically improved.

The *modus operandi* is to combine the stability of interpolating sequences with the stability of frames.

Theorem 13. *If the system $\{e^{i\lambda_n t}\}$ is a frame in $L^2[-\pi, \pi]$, then there is a positive constant L with the property that $\{e^{i\mu_n t}\}$ is also a frame in $L^2[-\pi, \pi]$ whenever $|\lambda_n - \mu_n| \leqq L$ for every n.*

Proof. Let $\{e^{i\lambda_n t}\}$ be a frame in $L^2[-\pi, \pi]$, with lower and upper bounds A and B, respectively. Then

$$A\|f\|^2 \leqq \sum_n |f(\lambda_n)|^2 \leqq B\|f\|^2$$

for every function f belonging to the Paley–Wiener space $\boldsymbol{P}$. Let L be a positive number and $\mu_1, \mu_2, \mu_3, \ldots$ complex scalars for which $|\lambda_n - \mu_n| \leqq L$ $(n = 1, 2, 3, \ldots)$. It is to be shown that if L is sufficiently small, then similar inequalities hold for the μ_n's.

By virtue of Lemma 3, for every f in $\boldsymbol{P}$,

$$\sum_n |f(\lambda_n) - f(\mu_n)|^2 \leqq B(e^{\pi L} - 1)^2\|f\|^2 \leqq C\sum_n |f(\lambda_n)|^2,$$

where

$$C = \frac{B}{A}(e^{\pi L} - 1)^2.$$

Applying Minkowski's inequality, we find that

$$\left|\sqrt{\sum |f(\lambda_n)|^2} - \sqrt{\sum |f(\mu_n)|^2}\right| \leqq \sqrt{C\sum |f(\lambda_n)|^2},$$

and hence that

$$\sqrt{A}(1 - \sqrt{C})\|f\| \leqq \sqrt{\sum |f(\mu_n)|^2} \leqq \sqrt{B}(1 + \sqrt{C})\|f\|$$

for every f. Since C is less than 1 if L is sufficiently small, the system $\{e^{i\mu_n t}\}$ is a frame in $L^2[-\pi, \pi]$. ■

Corollary. *If the system $\{e^{i\lambda_n t}\}$ is a Riesz basis for $L^2[-\pi, \pi]$ then there is a positive constant L with the property that $\{e^{i\mu_n t}\}$ is also a Riesz basis for $L^2[-\pi, \pi]$ whenever $|\lambda_n - \mu_n| \leqq L$ for every n.*

Proof. Let $\{e^{i\lambda_n t}\}$ be a Riesz basis for $L^2[-\pi, \pi]$ and suppose that $\mu_1, \mu_2, \mu_3, \ldots$ are complex scalars for which $\sup_n |\lambda_n - \mu_n| = L < \infty$. If L is sufficiently small, then (1) $\{\mu_n\}$ is an interpolating sequence for $\boldsymbol{P}$ (by Theorem 11) and (2) $\{e^{i\mu_n t}\}$ is a frame in $L^2[-\pi, \pi]$ (by Theorem 13) and hence complete. Now invoke Theorem 9. ■

The criterion of Theorem 13 is far too restrictive—the essential ingredient is not that $|\lambda_n - \mu_n|$ is small but rather that $\text{Re}(\lambda_n - \mu_n)$ is small. More precisely, we have the following result (cf. Theorem 3.12).

Theorem 14. *Let $\{\lambda_1, \lambda_2, \lambda_3, \ldots\}$ be a sequence of points lying in a strip parallel to the real axis. If the system $\{e^{i(\text{Re}\,\lambda_n)t}\}$ is a frame in $L^2[-\pi, \pi]$, then so is $\{e^{i\lambda_n t}\}$.*

Proof. Write $\lambda_n = \alpha_n + i\beta_n$, with α_n and β_n real, and choose M so large that $|\beta_n| \leqq M$ for every n. Suppose that the system $\{e^{i\alpha_n t}\}$ is a frame in $L^2[-\pi, \pi]$, with lower and upper bounds A and B, respectively. Then

$$A\|f\|^2 \leqq \sum_n |f(\alpha_n)|^2 \leqq B\|f\|^2$$

for every function $f \in \boldsymbol{P}$. We shall establish similar inequalities for the λ_n's. One half is easy. Since $|\lambda_n - \alpha_n| \leqq M$, it follows from Lemma 3 that

$$\sum_n |f(\lambda_n) - f(\alpha_n)|^2 \leqq B(e^{\pi M} - 1)^2\|f\|^2$$

for every f. A straightforward application of Minkowski's inequality then shows that

$$\sum_n |f(\lambda_n)|^2 \leqq Be^{2\pi M}\|f\|^2.$$

This establishes one half of the frame condition. It is convenient to prove the other half in three steps.

(i) Let $f(z)$ be a nontrivial function in $\boldsymbol{P}$. By Hadamard's factorization theorem

$$f(z) = z^k e^{az+b} \prod_{n=1}^{\infty} \left(1 - \frac{z}{z_n}\right) e^{z/z_n}, \tag{1}$$

where $z_1, z_2, z_3, \ldots$ are the zeros of $f(z)$ other than $z = 0$, k is the order of the zero at the origin, and a and b are complex numbers.

Since $f(z)$ is bounded on the real axis, the series

$$\sum_{n=1}^{\infty} \operatorname{Im}\left(\frac{1}{z_n}\right)$$

is absolutely convergent by Theorem 2.14. If $1/z_n = a_n + ib_n$, where a_n and b_n are real, then

$$f(z) = z^k e^{(c+id)z+b} \prod_{n=1}^{\infty}\left(1 - \frac{z}{z_n}\right) e^{a_n z}, \tag{2}$$

where $c = \operatorname{Re}(a)$ and $d = \operatorname{Im}(a) + \sum_{n=1}^{\infty} b_n$. Since the product in (1) converges uniformly on each compact set, so does the product in (2).

(ii) There exists a sequence $\lambda_1^{(1)}, \lambda_2^{(1)}, \lambda_3^{(1)}, \ldots$ of complex scalars and an entire function $f_1(z)$ belonging to $\boldsymbol{P}$ such that

$$\operatorname{Re}\lambda_n^{(1)} = \operatorname{Re}\lambda_n, \quad |\operatorname{Im}\lambda_n^{(1)}| \leqq \frac{M}{2} \quad (n = 1, 2, 3, \ldots) \tag{3}$$

and

$$\frac{\sum |f_1(\lambda_n^{(1)})|^2}{\|f_1\|^2} \leqq e^{\pi M} \frac{\sum |f(\lambda_n)|^2}{\|f\|^2}. \tag{4}$$

Assume that $d \geqq 0$. (The case $d < 0$ is similar and is left to the reader.) We begin by defining a function $g(z)$ whose zeros are obtained by reflecting across the real axis those zeros of $f(z)$ that lie in the lower half-plane. Specifically, we set

$$w_n = \begin{cases} z_n & \text{if } \operatorname{Im} z_n \geqq 0, \\ \bar{z}_n & \text{if } \operatorname{Im} z_n < 0, \end{cases}$$

and

$$g(z) = z^k e^{(c+id)z+b} \prod_{n=1}^{\infty}\left(1 - \frac{z}{w_n}\right) e^{a_n z}. \tag{5}$$

Assertion: $g(z) \in \boldsymbol{P}$ and

$$|g(x)| = |f(x)| \quad \text{whenever } x \text{ is real.} \tag{6}$$

To prove the assertion, define a sequence $g_1(z), g_2(z), g_3(z), \ldots$ of entire functions by writing

$$g_m(z) = f(z) \prod_{n=1}^{m} \frac{1 - z/w_n}{1 - z/z_n}, \quad m = 1, 2, 3, \ldots.$$

It is clear that $g_m(z) \to g(z)$ as $m \to \infty$, uniformly on every compact set, and hence $g(z)$ is entire. It is also clear that $g_m(z)$ and $f(z)$ have the same modulus on the real axis, and this establishes (6). It remains only to show that $g(z)$ is of exponential type at most π.

For each positive integer m the quotient $g_m(z)/f(z)$ approaches 1 as $|z| \to \infty$, and hence

$$|g_m(z)| \leqq \|f\| e^{\pi|z|}$$

for all z. Fixing z and letting $m \to \infty$, we see that $g(z)$ is an entire function of exponential type at most π. This proves the assertion.

Observe now that

$$|g(z)| \leqq |f(z)| \quad \text{whenever} \quad \operatorname{Im} z \geqq 0$$

(compare formulas (2) and (5)) and

$$|g(z)| \leqq |f(\overline{z})| \quad \text{whenever} \quad \operatorname{Im} z \geqq 0$$

since $d \geqq 0$. It follows that if

$$\mu_n = \begin{cases} \lambda_n & \text{if } \operatorname{Im} \lambda_n \geqq 0, \\ \overline{\lambda}_n & \text{if } \operatorname{Im} \lambda_n < 0, \end{cases}$$

then

$$|g(\mu_n)| \leqq |f(\lambda_n)| \quad (n = 1, 2, 3, \ldots).$$

Finally, define

$$\lambda_n^{(1)} = \mu_n - \frac{iM}{2} \quad (n = 1, 2, 3, \ldots) \quad \text{and} \quad f_1(z) = g\left(z + \frac{iM}{2}\right).$$

Then (3) holds and $|f_1(\lambda_n^{(1)})| \leqq |f(\lambda_n)|$ for every n. Thus, to establish (4), we have only to show that

$$\|f_1\|^2 \geqq e^{-\pi M} \|f\|^2.$$

Since $g(z) \in \boldsymbol{P}$, there is a function $\phi \in L^2[-\pi, \pi]$ such that

$$g(z) = \frac{1}{2\pi} \int_{-\pi}^{\pi} \phi(t) e^{izt}\, dt.$$

Since $f(z)$ and $g(z)$ have the same modulus on the real axis, we have

$$\int_{-\infty}^{\infty} |f(x)|^2\, dx = \int_{-\infty}^{\infty} |g(x)|^2\, dx = \frac{1}{2\pi} \int_{-\pi}^{\pi} |\phi(t)|^2\, dt,$$

by the Plancherel theorem. If we now set $\phi_1(t) = \phi(t) e^{-Mt/2}$, then

$$f_1(z) = \frac{1}{2\pi} \int_{-\pi}^{\pi} \phi_1(t) e^{izt}\, dt,$$

and hence

$$\int_{-\infty}^{\infty} |f_1(x)|^2\, dx = \frac{1}{2\pi} \int_{-\pi}^{\pi} |\phi(t)|^2 e^{-Mt}\, dt \geqq e^{-\pi M} \int_{-\infty}^{\infty} |f(x)|^2\, dx.$$

Thus $\|f_1\|^2 \geqq e^{-\pi M} \|f\|^2$.

(iii) We have completed the first step of an obvious induction procedure. Accordingly, given f, we can find for each positive integer k a sequence $\lambda_1^{(k)}, \lambda_2^{(k)}, \lambda_3^{(k)}, \ldots$ of scalars and an entire function $f_k(z)$ belonging to $\boldsymbol{P}$ such that

$$\operatorname{Re} \lambda_n^{(k)} = \operatorname{Re} \lambda_n, \quad \left|\operatorname{Im} \lambda_n^{(k)}\right| \leqq \frac{M}{2^k} \quad (n = 1, 2, 3, \ldots)$$

and

$$\frac{\sum_n |f_k(\lambda_n^{(k)})|^2}{\|f_k\|^2} \leqq e^{2\pi M} \frac{\sum_n |f(\lambda_n)|^2}{\|f\|^2}. \tag{7}$$

(The constant $2\pi M$ has been written in place of $\pi M + \frac{1}{2}\pi M + \cdots + \left(\frac{1}{2}\right)^{k-1} \pi M$.) Since the set $\{e^{i\alpha_n t}\}$ is a frame, there is a positive constant L such that $\{e^{i\gamma_n t}\}$ is also a frame whenever $|\alpha_n - \gamma_n| \leqq L$ by Theorem 13. Choose k so large that $M/2^k \leqq L$. Then the set $\{e^{i\lambda_n^{(k)} t}\}$ is a frame, and the left side of (7) has a positive lower bound A_1, independent of f. Hence

$$A_1 e^{-2\pi M} \|f\|^2 \leqq \sum_n |f(\lambda_n)|^2$$

for every $f \in \boldsymbol{P}$. ■

Corollary 1. *Let $\{\lambda_1, \lambda_2, \lambda_3, \ldots\}$ be a sequence of points lying in a strip parallel to the real axis. If the system $\{e^{i(\operatorname{Re} \lambda_n)t}\}$ is a Riesz basis for $L^2[-\pi, \pi]$, then so is $\{e^{i\lambda_n t}\}$*

Proof. It is sufficient to prove that the system $\{e^{i\lambda_n t}\}$ is an exact frame in $L^2[-\pi, \pi]$. That it is a frame follows from Theorem 14 since $\{e^{i(\operatorname{Re} \lambda_n)t}\}$ is a frame. Suppose then that a single arbitrary term $e^{i\lambda_k t}$ is deleted. Assertion: $\{e^{i(\operatorname{Re} \lambda_n)t}\}_{n \neq k}$ is incomplete in $L^2[-\pi, \pi]$. Reason: The removal of a vector from a frame leaves either a frame or an incomplete set. Conclusion: $\{e^{i\lambda_n t}\}_{n \neq k}$ is incomplete in $L^2[-\pi, \pi]$ by virtue of Theorem 3.12 and so cannot possibly be a frame. Thus the system $\{e^{i\lambda_n t}\}$ is an exact frame. ■

The following impressive generalization of Kadec's $\frac{1}{4}$-theorem is now immediate.

Corollary 2. *If $\{\lambda_n\}_{-\infty}^{\infty}$ is a sequence of scalars for which*

$$\sup_n |\operatorname{Re} \lambda_n - n| < \tfrac{1}{4} \quad \textit{and} \quad \sup_n |\operatorname{Im} \lambda_n| < \infty,$$

then the system $\{e^{i\lambda_n t}\}_{-\infty}^{\infty}$ is a Riesz basis for $L^2[-\pi, \pi]$.

More than 40 years after Paley and Wiener initiated the study of nonharmonic Fourier series in $L^2[-\pi, \pi]$, a complete solution to the exponential Riesz basis problem was found (Pavlov [1979]). For Pavlov's remarkable characterization, as well as recent developments, see the *Notes*.

Every example of a Schauder basis $\{e^{i\lambda_n t}\}$ for $L^2[-\pi, \pi]$ seen so far was proved to be a *Riesz basis*. In a sense this is not surprising—the class of Riesz bases is very large. Question: Are there bases of complex exponentials with purely imaginary exponents that are *not* Riesz bases? An example has not yet been found.

PROBLEMS

1. How must the proof of Theorem 14 be modified when $d < 0$?

2. Let $\{e^{i\lambda_n t}\}$ be a basis for $L^2[-\pi, \pi]$ and suppose that $\sup_n |\operatorname{Im} \lambda_n| < \infty$. Prove that $\{e^{i\mu_n t}\}$ is also a basis for $L^2[-\pi, \pi]$ provided that

$$\sum_{n=1}^{\infty} |\lambda_n - \mu_n| < \infty.$$

(*Hint*: By virtue of Theorem 1.12, it is sufficient to prove that

$$\sum_{n=1}^{\infty} \|e^{i\lambda_n t} - e^{i\mu_n t}\| < \infty \quad \text{whenever} \quad \sum_{n=1}^{\infty} |\lambda_n - \mu_n| < \infty.)$$

9 POINTWISE CONVERGENCE

The time has come to consider pointwise convergence. With regard to ordinary Fourier series, the deepest and most striking result is surely Carleson's proof of Lusin's conjecture: the Fourier series of every L^2 function converges pointwise almost everywhere. The following general "equiconvergence" theorem will show that nonharmonic Fourier series in L^2 have to a large extent the same convergence and summability properties as ordinary Fourier series.

Recall to begin with that two series $\sum a_n$ and $\sum b_n$ are said to be *equiconvergent* if their difference $\sum(a_n - b_n)$ converges and has the sum 0.

Theorem 15. *Let $\{e^{i\lambda_n t}\}_{-\infty}^{\infty}$ be a Riesz basis for $L^2[-\pi, \pi]$ and suppose that*

$$\sup_n |\lambda_n - n| < \infty.$$

For each function in $L^2[-\pi, \pi]$ the ordinary Fourier series and the nonharmonic Fourier series are uniformly equiconvergent on every compact subset of $(-\pi, \pi)$.

Proof. Let f be an arbitrary element of $L^2[-\pi, \pi]$ and δ a small positive number. It is to be shown that if f has the two norm-convergent expansions $\sum a_n e^{int}$ and $\sum c_n e^{i\lambda_n t}$, then the difference

$$\sum_{n=-N}^{N} (a_n e^{int} - c_n e^{i\lambda_n t})$$

converges to zero as $N \to \infty$, uniformly on the interval $[-\pi + \delta, \pi - \delta]$. The essence of the proof is to obtain a suitable representation for this difference.

Begin by writing

$$e^{i\lambda_n t} = e^{int} e^{i(\lambda_n - n)t} = e^{int} \sum_{k=0}^{\infty} b_{nk} t^k,$$

where

$$b_{nk} = \frac{i^k (\lambda_n - n)^k}{k!} \qquad (k \geqq 0, -\infty < n < \infty).$$

If f_N denotes the Nth partial sum of the nonharmonic Fourier series of f and

$$\psi_{Nk}(t) = \sum_{n=-N}^{N} c_n b_{nk} e^{int},$$

then

$$f_N(t) = \sum_{k=0}^{\infty} t^k \psi_{Nk}(t) \quad \text{for} \quad |t| \leqq \pi \quad (N = 1, 2, 3, \ldots).$$

Choose L so large that $|\lambda_n - n| \leqq L$ for every n. Then

$$\begin{aligned} ||\psi_{Nk}||^2 &= \sum_{n=-N}^{N} |c_n b_{nk}|^2 \quad \text{(by Parseval's identity)} \\ &\leqq \frac{L^{2k}}{(k!)^2} \sum_{n=-N}^{N} |c_n|^2, \end{aligned}$$

and hence

$$\psi_k \equiv \underset{N\to\infty}{\text{l.i.m.}}\ \psi_{Nk}$$

exists for every k. (Here l.i.m. means *limit in mean* in the L^2 sense.) Observe that $\sum_{n=-\infty}^{\infty} |c_n|^2 = c < \infty$ and $||\psi_k|| \leqq cL^k/k!$ for $k \geqq 0$.

Assertion:

$$f(t) = \sum_{k=0}^{\infty} t^k \psi_k(t) \quad \textit{(in the mean)}.$$

Indeed, since $\sum \left\|t^k \psi_k(t)\right\| \leqq c \sum (\pi L)^k/k! < \infty$, the series $\sum t^k \psi_k(t)$ is absolutely convergent and so must converge in $L^2[-\pi, \pi]$ to some element g of $L^2[-\pi, \pi]$. We have

$$\begin{aligned} \|g - f_N\| &\leqq \sum_{k=0}^{\infty} \left\|t^k(\psi_k - \psi_{Nk})\right\| \\ &\leqq \sum_{k=0}^{\infty} \pi^k \, \|\psi_k - \psi_{Nk}\| \leqq 2c \sum_{k=0}^{\infty} \frac{(\pi L)^k}{k!} < \infty, \end{aligned}$$

and since $\lim_N \|\psi_k - \psi_{Nk}\| = 0$ for each k, we must have $\lim_N \|g - f_N\| = 0$. But $\lim_N \|f - f_N\| = 0$, and therefore $f(t) = g(t)$ almost everywhere on $[-\pi, \pi]$. This proves the assertion.

Let D_N be the *Dirichlet kernel*†, i.e.,

$$D_N(t) = \frac{\sin\left(N + \frac{1}{2}\right)t}{\sin \frac{1}{2}t} \qquad (N = 1, 2, 3, \ldots),$$

and g_N the Nth partial sum of the ordinary Fourier series of f. Then

$$g_N(t) = (f(x), D_N(x - t)) = \sum_{k=0}^{\infty} (x^k \psi_k(x), D_N(x - t)).$$

It follows at once from the definitions of ψ_{Nk} and ψ_k that

$$f_N(t) = \sum_{k=0}^{\infty} t^k (\psi_k(x), D_N(x - t)),$$

and hence

$$\sum_{n=-N}^{N} (a_n e^{int} - c_n e^{i\lambda_n t}) = \sum_{k=0}^{\infty} (\psi_k(x), (x^k - t^k) D_N(x - t))$$

whenever $|t| \leqq \pi$. This is the desired representation.

We complete the proof by estimating the inner products

$$(\psi_k(x), (x^k - t^k) D_N(x - t)).$$

Denote by Ω the closed rectangle $\{(x, t)\colon |x| \leqq \pi, |t| \leqq \pi - \delta\}$, and by H_k the function

$$H_k(x, t) = \frac{x^k - t^k}{\sin \frac{1}{2}(x - t)} \quad \text{for} \quad (x, t) \in \Omega \quad (k = 0, 1, 2, \ldots).$$

† To simplify the notation, we have suppressed a factor of $\frac{1}{2}$.

It is clear that each H_k is continuous on Ω. It is not difficult to show (the proof is left to the reader) that there is a constant A such that

$$|H_k(x,t)| \leqq A\pi^k \quad \text{for} \quad (x,t) \in \Omega \quad (k = 0, 1, 2, \ldots).$$

Therefore, if $|t| \leqq \pi - \delta$, then

$$|(\psi_k(x), (x^k - t^k)D_N(x - t))| \leqq A\pi^k \,||\psi_k|| \leqq cA\frac{(\pi L)^k}{k!}$$

for all values of N and k. Given $\varepsilon > 0$, choose M so large that

$$\sum_{k>M} \frac{(\pi L)^k}{k!} < \frac{\varepsilon}{cA}.$$

Then, for every N,

$$\sum_{k>M} |(\psi_k(x), (x^k - t^k)D_N(x - t))| < \varepsilon \quad \text{whenever} \quad |t| \leqq \pi - \delta.$$

Since H_k is uniformly continuous on Ω, a simple extension of the Riemann-Lebesgue lemma shows that

$$\sum_{k \leqq M} |(\psi_k(x), (x^k - t^k)D_N(x - t))| \to 0 \quad \text{as} \quad N \to \infty,$$

uniformly for $|t| \leqq \pi - \delta$. This proves that

$$\lim_{N\to\infty} \sum_{-N}^{N} (a_n e^{int} - c_n e^{i\lambda_n t}) = 0$$

uniformly on each compact subset of $(-\pi, \pi)$. ∎

We conclude that the nonharmonic Fourier series of an L^2 function converges at a given point of $(-\pi, \pi)$ if and only if the ordinary Fourier series converges at that point; by Carleson's theorem, this occurs almost everywhere. Furthermore, on every compact subset of $(-\pi, \pi)$, the summability properties of the two series are uniformly the same (Problem 2).

PROBLEMS

1. Let H_k and Ω be defined as in Theorem 15. Show that there is a constant A such that

$$|H_k(x,t)| \leqq A\pi^k \quad \text{for} \quad (x,t) \in \Omega \quad (k = 0, 1, 2, \ldots).$$

2. Prove that Theorem 15 remains valid when the term "equiconvergent" is replaced by "equisummable", for any summability method that is both *regular* and *linear*. (For an excellent treatment of such methods, which include $(C, 1)$, Abel, and Riesz summability, see Szász [1944].)

3. Let $\{\lambda_n\}_{-\infty}^{\infty}$ be a sequence of complex numbers for which

$$\sup_n |\operatorname{Re}\lambda_n - n| < \tfrac{1}{4} \quad \text{and} \quad \sup_n |\operatorname{Im}\lambda_n| < \infty.$$

Prove that the system $\{e^{i\lambda_n t}\}_{-\infty}^{\infty}$ is complete in $C(I)$, for each closed subinterval I of $(-\pi, \pi)$.

4. By modifying the proof of Theorem 15, show that

$$(\pi - |t|)\sum_{-N}^{N}(a_n e^{int} - c_n e^{i\lambda_n t}) \to 0 \quad \text{as} \quad N \to \infty,$$

uniformly on $[-\pi, \pi]$.

NOTES AND COMMENTS

CHAPTER 1

SECTION 1

General references. The first systematic investigation of bases in Banach spaces was made by Banach [1932]. Since then, research in this area has flourished. Two excellent references, the first an introductory text, the second an encyclopedic account of bases in Banach spaces, are by Marti [1969] and Singer [1970]. The important work of Lindenstrauss and Tzafriri [1977] deals mainly with recent results and current research directions.

Weak bases. The definition of a basis for a Banach space can be weakened. The "weak basis theorem" asserts that a *weak basis* for a Banach space, i.e., a basis relative to the weak topology, is in fact a Schauder basis. This is stated without proof in Banach [1932, p. 238]. Karlin [1948] sketches a proof. A complete proof, valid for every Fréchet space, is given in Bessaga and Pelczynski [1959].

Bases for special spaces. Most of the common Banach spaces are now known to possess bases. Among the many important examples, the following deserve special mention. In $L^p[-\pi, \pi]$, $1 < p < \infty$, the *trigonometric system* $\{e^{int}\}_{-\infty}^{\infty}$ and in $L^p[0, 1]$, $1 \leqq p < \infty$, the *Haar system* $\{h_n\}$ constitute

Schauder bases. By definition, $h_1(t) \equiv 1$ and

$$h_{2^k+m}(t) = \begin{cases} 1, & t \in [(2m-2)/2^{k+1}, (2m-1)/2^{k+1}) \\ -1, & t \in [(2m-1)/2^{k+1}, 2m/2^{k+1}) \\ 0, & \text{elsewhere} \end{cases}$$

$(k = 0, 1, 2, \ldots, m = 1, \ldots, 2^k)$ (see Marti [1969, p. 49]). If we put

$$\phi_1(t) \equiv 1 \quad \text{and} \quad \phi_{n+1}(t) = \int_0^t h_n(u)\,du \quad \text{for} \quad 0 \leqq t \leqq 1 \ (n = 1, 2, 3, \ldots),$$

then we obtain the *Schauder system* $\{\phi_n\}$; it is a basis for $C[0, 1]$ (see Section 2). More recently, Billard [1972] has constructed a basis for the Hardy space H^1; Ciesielski and Domsta [1972] and Schonefeld [1972] have independently constructed a basis for $C^k(I^n)$; and Bočkarev [1974] has constructed a basis for the disk algebra A, using the *Franklin system* in an essential way. (The Franklin system is obtained from the Schauder system by applying the Gram-Schmidt orthogonalization procedure. It is also a basis for $C[0, 1]$. For a detailed study of the Franklin system, see Ciesielski [1963, 1966].) For further examples see Pelczynski [1977].

Problem 8. It was shown by Riemann that in a finite-dimensional Banach space the class of absolutely convergent series and the class of unconditionally convergent series coincide. In an infinite-dimensional Banach space, they do not. **Theorem (Dvoretzky and Rogers [1950]).** *If X is an infinite-dimensional Banach space and if $\{c_n\}$ is an arbitrary square-summable sequence of positive numbers, then there are elements $x_n \in X$ such that $\|x_n\| = c_n$ $(n = 1, 2, 3, \ldots)$ and the series $\sum_{n=1}^{\infty} x_n$ is unconditionally convergent.*

Conditional and unconditional bases. A basis $\{x_n\}$ for a Banach space is said to be *unconditional* if every convergent series of the form $\sum_{n=1}^{\infty} c_n x_n$ is unconditionally convergent. The trigonometric system is unconditional in $L^p[-\pi, \pi]$ only when $p = 2$ (Karlin [1948]). The following two systems are both unconditional in $L^p[0, 1]$ for $1 < p < \infty$: the Haar system (Marcinkiewicz [1937]) and the Franklin system (Ciesielski, Simon, and Sjölin [1977]). Since $L^p[0, 1]$ and H^p are isomorphic for $1 < p < \infty$ (Boas [1955]), it follows that $H^p (1 < p < \infty)$ also has an unconditional basis. For an elegant proof, based on wavelets, that H^1 also has an unconditional basis, see Meyer [1992, Chapter 6]. (For an excellent account of H^p-spaces, see Hoffman [1962] and Duren [1970].) Wavelets are now known to provide unconditional bases for most of the classical function spaces (see Daubechies [1992] and Meyer [1992], and, for an elementary exposition, Hernández and Weiss [1996]). There are also spaces that admit no unconditional bases: $C[0, 1]$ (Karlin [1948]) and $L^1[0, 1]$ (Pelczynski [1961]), for example (cf. Arutjunjan [1972]). In this connection the following result due to Krotov [1974] is striking: *There exists a continuous function on* $[0, 1]$ *whose expansion in the Schauder system, with a suitable ordering of the terms, converges almost everywhere to any prescribed measurable function.* (See Goffman [1964] for a related result.) The first example of a reflexive separable Banach space that admits no unconditional basis was given by Kwapien and Pelczynski [1970] (see also Lindenstrauss and Pelczynski [1971]). Since there are spaces in which all bases are conditional, it is natural to ask if there are spaces in which all bases are unconditional. The answer is no—every Banach space with a basis has a conditional basis (Pelczynski and Singer [1964]; see the *Notes* for Section 8).

The approximation property. A Banach space X is said to have the *approximation property* (in the sense of Grothendieck [1955]) if the identity operator on X can be approximated, uniformly on every compact subset of X, by operators of finite rank. Enflo's example provides a separable Banach space that fails to have the approximation property and so a fortiori fails to have a basis (see Problem 7, Section 6). The example has been greatly simplified by Davie [1973], who shows that the spaces c_0 and $l^p (2 < p < \infty)$ have subspaces that fail to have the approximation property. It is worth noting that it is usually much easier to verify that a given space has the approximation property than to show that it has a basis. Thus, for example, it is relatively simple to show that the disk algebra A has the approximation property (see Problem 8, Section 6), while the existence of a basis for A was for many years an open problem.

SECTION 2

The Weierstrass approximation theorem and related theorems. The classical theorem of Weierstrass on the approximation of continuous functions by polynomials appears in Weierstrass [1885]. Bernstein's proof, which is based on probability theory, is from Bernstein [1912a]. (For a penetrating study of the Bernstein polynomials, see Lorentz [1953].) Weierstrass's theorem has been generalized in many different directions. The important extensions of Stone, Müntz, and Mergelyan are well known (see, e.g., Rudin [1966, pp. 305, 386] and Royden [1968, p. 174]). We mention, in addition, the following result of Wolibner [1951] on simultaneous interpolation and approximation: *If* $f \in C[a, b]$ *and if* $\{x_1, \ldots, x_n\}$ *is a set of n arbitrary points in the interval* $[a, b]$, *then for each* $\varepsilon > 0$ *there is a polynomial p such that* (1) $\|f - p\| < \varepsilon$, (2) $\|p\| = \|f\|$, *and* (3) $p(x_i) = f(x_i)$, $i = 1, \ldots, n$. (This generalizes a well-known result of Walsh [1935, p. 310] in which properties (1) and (3) were shown to hold.) In Deutsch and Morris [1969], property SAIN (simultaneous approximation and interpolation which is norm-preserving) is introduced and studied in an abstract normed vector space.

Theorem 1. Schauder [1927]. Meletidi [1972] has shown that for arbitrary positive numbers A and $B\,(A < B)$ there exists a basis $\{f_n\}$ for $C[0, 1]$ such that $A \leqq f_n(t) \leqq B$ for $n = 1, 2, 3, \ldots$ and all values of $t \in [0, 1]$.

SECTION 3

General references. For Hilbert space theory, Halmos [1951] and Berberian [1961] are elementary references; for a more advanced treatment, see Riesz and Sz.-Nagy [1955], Dunford and Schwartz [1958, 1963], and Halmos [1967]. For H^p-spaces, see Hoffman [1962] and Duren [1970]. And for Fourier series, Zygmund [1959] is the standard work; for a more elementary introduction, see Titchmarsh [1939].

Problem 6. This definition of orthogonality was introduced by Birkhoff [1935] and extensively developed by James [1945].

Problem 8. The theory of almost periodic functions, an elegant and natural generalization of the theory of periodic functions, was created by Bohr [1925a, 1925b, 1926]. (For an excellent account of the elementary theory, see Bohr [1947].) Since then it has been considerably extended (see, e.g., Stepanoff [1925], Besicovitch [1932], and Paley and Wiener [1934, p. 116]; other references may be found in Dunford and Schwartz [1958]).

Problem 12. Vitali [1921].

Problem 13. Dalzell [1945].

Problem 14. For an excellent account of the classical orthogonal polynomials, see Szegö [1939].

Problem 16. These facts were first discussed by Bergman [1947, p. 24]. The suggested proof is from Halmos [1967, p. 187].

SECTION 4

General references. Aronszajn [1950] and Hille [1972]. Additional references may be found in Shapiro [1971, p. 101]. For a deeper discussion of the Bergman kernel and its application to conformal mappings, see Bergman [1950] and Nehari [1952].

SECTION 5

Problem 6. Boas and Pollard [1948].

Problem 8. A classical treatise on closure theorems for analytic functions is Walsh [1935] (see also Davis [1963]).

SECTION 6

Theorem 3. Banach [1932, p. 111]. Schauder's original definition of a basis had required the continuity of the coefficient functionals.

Theorem 4. James [1950].

Problem 5. The term "normal basis" was introduced by Karlin [1948], although the concept already appears in Banach [1932, p. 238]. The suggested proof is from Diestel [1975, p. 43]. For another simple proof, see Cook [1970].

Problem 7. For a detailed investigation of the many variants of the approximation property and the relations between them, see Grothendieck [1955].

SECTION 7

Bases for the dual space. Let X be a Banach space with a basis. Even if X^* is separable, it need not have a basis. This is hard—the proof makes use of the existence of a separable Banach space without a basis (see Lindenstrauss and Tzafriri [1977, p. 34]). A deep result of Johnson, Rosenthal, and Zippin [1971] states that if X^* has a basis, then so does X. (The problem had been posed by Karlin [1948].)

Theorem 5. This result is essentially due to Banach [1932, p. 107].

Problem 3. Karlin [1948].

Problem 4. In fact, for each infinite-dimensional closed subspace K of a Hilbert space H, there exist biorthogonal sequences $\{f_n\}$ and $\{g_n\}$ such that $\{f_n\}$ is complete in H and the closed subspace spanned by $\{g_n\}$ is K. For minimal sequences of complex exponentials in $L^2[-\pi, \pi]$ the situation is dramatically different—*the completeness of such a sequence always ensures the completeness of its biorthogonal sequence* (Young [1981]). This result was used in a novel way by Gröchenig, Heil, and Walnut [2000] to prove the *local three squares theorem*.

SECTION 8

Equivalent bases. Equivalent bases were studied by Arsove [1958] (he calls them "similar bases"). Theorem 7 was proved by Arsove [1958] for locally convex Fréchet spaces and, independently, by Bessaga and Pelczynski [1958]. It has recently been shown by Ciesielski, Simon, and Sjölin [1977] that the Haar and Franklin systems are equivalent in $L^p[0, 1]$, $1 < p < \infty$. Sjölin [1977] has shown that this result is not true if $p = 1$.

Uniqueness of bases. Once a Banach space is known to have a basis, it is natural to ask about its "uniqueness", up to equivalence. In a finite-dimensional Banach space all bases are equivalent. In an infinite-dimensional Banach space the situation is dramatically different—bases, if they exist at all, are never unique.

Theorem (Pelczynski and Singer [1964]). *In every infinite-dimensional Banach space with a basis, there exist uncountably many nonequivalent normalized conditional bases.*

(*Normalized* means that each basis element has norm 1.)

If a basis for a Hilbert space is to be equivalent to an orthonormal basis, i.e., if it is to be a Riesz basis, then it must be at the very least both bounded and unconditional. Remarkably, the converse is also true: *in a separable Hilbert space all bounded unconditional bases are equivalent.* This observation goes back to Köthe [1936] and Lorch [1939]. It was rediscovered by Gelbaum [1950] and Gelfand [1951]. Since then it has been reproved many times (see the references in Singer [1970, p. 640]). The quickest proof (see Lindenstrauss and Tzafriri [1977, p. 71]) makes use of Orlicz's theorem: *If $\sum x_n$ is an unconditionally convergent series in a Hilbert space, then $\sum \|x_n\|^2 < \infty$* (Orlicz [1933]; for a one-line proof of Orlicz's theorem, see Lindenstrauss and Tzafriri [1977, p. 18]). The spaces c_0 and l^1 also have the property that all bounded unconditional bases are equivalent (Lindenstrauss and Pelczynski [1968]). There are no other examples. The only infinite-dimensional Banach spaces with an unconditional basis, in which all normalized unconditional bases are equivalent, are (up to an isomorphism) c_0, l^1, and l^2 (Lindenstrauss and Zippin [1969]).

Theorem 9. Bari [1951].

Conditional bases for Hilbert space. The first example of a conditional basis for $L^2[-\pi, \pi]$ was given by Babenko [1948] (see also Gelbaum [1951] and Bari [1951]). For a simple example of a conditional basis for l^2, see Lin and Singer [1971]. Although we have dealt almost exclusively with unconditional bases, the following result deserves attention:

Theorem (Gurariĭ and Gurariĭ [1971]). *If $\{f_n\}$ is a normalized basis for a Hilbert space H, then there exist positive constants A, B, p, and q, with $1 < p \leqq 2 \leqq q < \infty$, such that for each convergent expansion $f = \sum c_n f_n$ we have*

$$A\left(\sum |c_n|^q\right)^{1/q} \leqq \|f\| \leqq B\left(\sum |c_n|^p\right)^{1/p}.$$

(For a different proof, see Ruckle [1972].)

SECTION 9

General references. Singer [1970] and Retherford and Holub [1971].

Theorem 10. The original form of this theorem, with $X = L^2[a, b]$ and $\{x_n\}$ an orthonormal basis for X, is due to Paley and Wiener [1934, p. 100]. The generalization given here is due to Boas [1940]. After further generalizations by Pollard [1944] and Sz.-Nagy [1947], Hilding [1948] gave the most general form of the theorem:

Theorem. *If X is a Banach space and $T : X \to X$ is a bounded linear transformation such that*

$$||(I - T)x|| \leq \lambda(||x|| + ||Tx||)$$

for some constant λ, $0 \leq \lambda < 1$, and all $x \in X$, then T is invertible.

For further results in this direction, see Singer [1970, pp. 84–109, 337–359] and Kato [1976, Chapter IV]. Generalizations in a different direction are given by Bourgin [1946], Gosselin and Neuwirth [1968/69], Davis [1969], and Casazza and Kalton [1999].

Corollary (Theorem 10) and Theorem 12. Krein and Liusternik [1948].

Theorem 11. Krein, Milman, and Rutman [1940]. An extension to locally convex spaces has been given by Lerer [1969]. The following result of Gurariĭ and Meletidi [1970] had been for many years an open problem. **Theorem.** *If $\{x_n\}$ is a complete sequence of elements from a Banach space X, then there exist positive numbers ε_n such that any sequence $\{y_n\}$ for which $\|x_n - y_n\| < \varepsilon_n$ $(n = 1, 2, 3, \ldots)$ is also complete in X.*

Problem 3. Retherford and Holub [1971].

SECTION 10

Nonharmonic Fourier series. The study of nonharmonic Fourier series was initiated by R. E. A. C. Paley and N. Wiener in their celebrated treatise of 1934. In this work the criterion of Theorem 13 is introduced and verified whenever each λ_n is real and $|\lambda_n - n| \leqq L < 1/\pi^2$ $(n = 0, \pm 1, \pm 2, \ldots)$. Ultimately, it was shown that the constant $1/\pi^2$ could be replaced by $\frac{1}{4}$ (see Theorem 14). That $\frac{1}{4}$ is in fact the "best possible" constant follows from a theorem of Levinson [1940, p. 48]: *The system of exponentials* $\{e^{\pm i(n-\frac{1}{4})t}$: $n = 1, 2, 3, \ldots\}$ *is complete in* $L^2[-\pi, \pi]$ (see Theorem 3.5). Young [1974c] has noted that the condition $|\lambda_n - n| < \frac{1}{4}$ $(n = 0, \pm 1, \pm 2, \ldots)$ is also insufficient for the system $\{e^{i\lambda_n t}\}$ to form a Riesz basis for $L^2[-\pi, \pi]$ (see also Young [1984, 1987]). The theory of nonharmonic Fourier series has found practical applications in such diverse areas as the theory of diffusion (see Hammersley [1953]), control theory (see Russell [1967], Fattorini and Russell [1974], Fattorini [1977], Avdonin and Ivanov [1984, 1995], Avdonin, Ivanov, and Io [1990], Hansen [1991], Avdonin, Ivanov, and Joó [1995], and Avdonin, Ivanov, and Russell [1999]), and the solution of the neutron slowing down equation (see Sengupta and Karnick [1978]). For additional applications see Freund and Petzold [1984], Narukawa and Suzuki [1986], and Ivanov [1996].

Theorem 14. Kadec [1964]. A striking generalization was discovered by Avdonin [1979]:

Avdonin's Theorem of "$\frac{1}{4}$ in the mean". *Let* $\lambda_n = n + \delta_n$ $(n = 0, \pm 1, \pm 2, \ldots)$ *be a separated sequence of real or complex numbers. If there exists a positive integer N and a constant d,* $0 \leqq d < \frac{1}{4}$, *such that*

$$\left| \sum_{k=mN+1}^{(m+1)N} \delta_k \right| \leq dN$$

for all integers m, then the system $\{e^{i\lambda_n t}\}$ *is a Riesz basis for* $L^2[-\pi, \pi]$.

For a more general statement see Avdonin and Ivanov [1995, p. 102].

Theorem 15. Bari [1951].

Theorem 16. Birkhoff and Rota [1960].

Sturm–Liouville series. The idea of expanding an arbitrary function in terms of the solutions of a second-order differential equation goes back to the time of C. Sturm and J. Liouville (see Titchmarsh [1946]). The proof of Theorem 17, based on Theorem 16, is due to Birkhoff and Rota [1960]. For a generalization of Theorem 16, together with an application to certain non-selfadjoint boundary value problems, see Brauer [1964] (see also Wallen [1969]).

Theorem 18 and Problems 1, 2. Duffin and Eachus [1942]. The suggested proof of Problem 2 is due to Riesz and Sz.-Nagy [1955, p. 209]. Jaffard and Young [1998] have extended Theorem 18 by showing that *every Schauder basis for a separable Hilbert space can be obtained from an orthonormal basis by means of a (possibly unbounded) one-to-one positive self adjoint operator.*

Problem 3. Schäfke [1949]. The suggested proof is from Retherford and Holub [1971].

Problem 4. Birkhoff and Rota [1960]. A somewhat simpler argument may be found in Halmos [1967, Problem 7].

CHAPTER 2

General references. Borel [1921], Valiron [1949], Boas [1954], Cartwright [1956], Levin [1964], Markushevich [1966], and Holland [1973]. See also Titchmarsh [1939] and Markushevich [1965].

PART ONE

SECTION 1

Information on infinite products in general can be found in Knopp [1951].

Theorem 1. Weierstrass [1876].

SECTION 2

Theorem 2. Jensen [1899]. For alternative proofs, see Titchmarsh [1939].

SECTION 3

The various possible relations between p, λ, and ρ are illustrated in Shah [1941].

SECTION 4

Theorem 6. Borel [1897].

Theorem 8. For more precise results, see the *Notes* for Section 5.

SECTION 5

Theorem 9. Hadamard [1893]. For another proof, see Titchmarsh [1939, p. 250].

The minimum modulus. Let $f(z)$ be a nonconstant entire function and let $m(r)$ be its minimum modulus for $|z| = r$. It follows readily from Hadamard's factorization theorem that if $f(z)$ is of order ρ, then for each $\varepsilon > 0$,

$$m(r) > e^{-r^{\rho+\varepsilon}}$$

on circles $|z| = r$ of arbitrarily large radius (see Problem 4). Under special conditions, much more can be said. For example, if $0 < \rho < 1$, then there are arbitrarily large values of r such that

$$m(r) > (M(r))^{\cos \pi\rho - \varepsilon}$$

(see Titchmarsh [1939, p. 275]). For further results, see Boas [1954] and Holland [1973].

Problem 11. See Titchmarsh [1939, p. 266].

PART TWO

SECTION 1

Theorem 10. Phragmén and Lindelöf [1908]. For additional theorems of this type, see Titchmarsh [1939].

Theorem 11. See, e.g., Pólya and Szegö [1925, Part IV, Problem 202], Duffin and Schaeffer [1938], and Plancherel and Pólya [1938]. If $f(z)$ is real for real z, then there is a sharper result:

$$|f(x+iy)| \leqq M \cosh By;$$

moreover, if equality holds for a single nonreal point, then $f(z) = M\cos(Bz + C)$, where C is real (Duffin and Schaeffer [1938]; for another proof, see Redheffer [1953]).

It is frequently possible to determine the growth of an entire function of exponential type along a line from its growth along a sequence of points on or near the line. For example, Cartwright [1935] has shown that an entire function of exponential type $< \pi$ that is bounded on the integers must be bounded on the entire real axis. This was generalized by Duffin and Schaeffer [1945], who showed that the integers can be replaced by any sequence having *uniform density* 1. (A sequence $\{\lambda_n\}_{-\infty}^{\infty}$ of real or complex numbers is said to have uniform density 1 if it is separated and $\sup_n |\lambda_n - n| < \infty$.) There are similar results when $\{f(\lambda_n)\} \in l^p$ (see the *Notes* for Section 3).

Bernstein's inequality. Bernstein proved the inequality named after him first for trigonometric polynomials (Bernstein [1912b]) and then later for arbitrary functions in $\boldsymbol{B}_\tau$ (Bernstein [1923]). The following generalization is due to Duffin and Schaeffer [1938] (see also Akhiezer [1956, p. 144]): If $f(z) \in \boldsymbol{B}_\tau$ and is real for real z, then

$$\sup_{-\infty<x<\infty} \sqrt{|f'(x)|^2 + \tau^2|f(x)|^2} \leqq \tau \sup_{-\infty<x<\infty} |f(x)|.$$

This shows not only that $|f'(x)| \leqq \tau \sup |f(x)|$ but also that $|f(x)|$ can be large only when $|f'(x)|$ is small. For further generalizations and improvements, see Boas [1954, pp. 210–218] and Giroux, Rahman, and Schmeisser [1979].

Problem 9. The suggested proof is due to Pólya and Szegö [1925, Part III, Problems 295, 298].

Problem 10. See Theorem 14. For yet another proof, see Luxemburg and Korevaar [1971, p. 26] (cf. Levinson [1940, p. 25]).

Problem 12. This proof is due to Boas [1954, p. 210].

Problem 13. For another simple and elegant proof, see O'Hara [1973].

SECTION 2

Theorem 13. Carleman [1922]. Here "Carleman's formula" is introduced for the purpose of proving Müntz's theorem.

Müntz's theorem (Müntz [1914], Szász [1916]). For a very good review of work on Müntz–Szász type approximation since N. Bernstein proposed the problem in 1912, see Luxemburg [1976]. Of the many proofs of Müntz's theorem that have been given, the following deserve attention: Pólya and Szegö [1925, Part IV, Problem 198], Paley and Wiener [1934, p. 32], Boas [1954, p. 235], Akhiezer [1956, p. 43], Rudin [1966, p. 305], Feller [1968], Luxemburg and Korevaar [1971], and Shapiro [1971, p. 99]. The converse is also true: if $\sum 1/\lambda_n < \infty$, then the system $\{t^{\lambda_n}\}$ is incomplete in $C[a, b]$. In this case a complete characterization of the closed linear span of the system $\{t^{\lambda_n}\}$ was provided by Clarkson and Erdös [1943], Korevaar [1947], and Schwartz [1959]. For an expository account of these and related results, see Anderson [1978].

Density. If $f(z)$ is an entire function of exponential type $\tau > 0$, bounded on the real axis, then the zeros of $f(z)$ in each half-plane $x > 0$ and $x < 0$ have density τ/π. (Levinson [1940, Chap. III] proves an even stronger result.) The first result of this sort appears to be due to Titchmarsh [1926, p. 285].

Problem 4. See Titchmarsh [1939, p. 132].

Problem 5. Titchmarsh [1926, p. 284; 1939, p. 137].

SECTION 3

All the results of this section are due to Plancherel and Pólya [1938]. See Boas [1954, pp. 97–103] for a wide range of generalizations.

Theorem 17. There is a variety of converse results when $f(z)$ is of sufficiently small type. If, for example, $f(z)$ is of exponential type $\tau < \pi$ and if for some positive number p, $\sum_{-\infty}^{\infty} |f(n)|^p < \infty$, then

$$\int_{-\infty}^{\infty} |f(x)|^p \, dx \leqq B \sum_{-\infty}^{\infty} |f(n)|^p,$$

where B depends only on p and τ (Plancherel and Pólya [1938]; see also Boas [1940]). An analogous result is valid if the integers are replaced by any sequence having uniform density 1 (Duffin and Schaeffer [1952] for $p = 2$; Boas [1952]).

Problem 2. See Boas [1954, p. 99]. Additional references may be found in Boas [1938].

Problem 4. For a sharper estimate, see Boas [1954, p. 102].

Problem 7. Plancherel and Pólya [1938]. When $p \geqq 1$, A can be taken to be τ^p (see, e.g., Boas [1954, p. 211]).

SECTION 4

Theorem 18. Paley and Wiener [1934, p. 12]. The proof given in the text is from Boas [1954, p. 105]. For additional proofs, see Boas [1954, p. 106], Rudin [1966, p. 371], and de Branges [1968, p. 47]. There are analogous Fourier representations for entire functions of exponential type that belong to L^p on the real axis, as well as for functions that are analytic and of exponential type in a half-plane and belong to L^p on the boundary (see, e.g., Boas [1954, p. 107] and Rudin [1966, p. 368]).

Example 2. This proof is due to Boas [1937].

Example 3. Boas [1972].

Problem 3. This problem appears in Boas [1954, p. 244].

SECTION 5

The Paley–Wiener space and other Hilbert spaces of entire functions are discussed from a different point of view in de Branges [1968].

For an extensive treatment of the cardinal series, see Whittaker [1935] and the excellent survey by Higgins [1985]. For further results on *irregular sampling*, see Benedetto [1992], Feichtinger and Gröchenig [1992, 1994], and Flornes, Lyubarskiĭ, and Seip [1999].

Problem 3. See Boas [1972] and Pollard and Shisha [1972].

Problem 4. See, e.g., Young [1974a, p. 112].

CHAPTER 3

General references. Paley and Wiener [1934, Chap. VI], Levinson [1940], Schwartz [1959], and especially Redheffer [1977].

SECTION 1

The terms *exact, excess*, and *deficiency* were introduced into mathematical terminology by Paley and Wiener [1934, p. 92].

Corollary (Theorem 1). The following alternative proof, based on the F. and M. Riesz theorem, was suggested by P. S. Muhly. If μ is a complex Borel measure on the interval $[-\pi, \pi]$ such that

$$\int_{-\pi}^{\pi} e^{int}\, d\mu(t) = 0 \quad (n = 1, 2, 3, \ldots),$$

then μ is absolutely continuous with respect to Lebesgue measure and $d\mu/dt \in H^1$. (As is customary, we have identified H^1 with the subspace of $L^1[-\pi, \pi]$ spanned by the exponentials e^{int} with $n \geqq 0$.) Since a function in H^1 cannot vanish on a set of positive measure unless it vanishes identically (see Hoffman [1962, p. 52]), the result follows.

Theorem 2. Let $\{\lambda_1, \lambda_2, \lambda_3, \ldots\}$ be an increasing sequence of positive numbers. Szász [1933], solving a problem of Pólya [1931], proved that if lim inf $n/\lambda_n > 1$, then the conditions

$$f \in C[-\pi, \pi] \quad \text{and} \quad \int_{-\pi}^{\pi} f(t) \sin \lambda_n t \, dt = \int_{-\pi}^{\pi} f(t) \cos \lambda_n t \, dt = 0$$

imply that f is identically zero. Under the weaker assumption lim sup $n/\lambda_n >$ 1, Paley and Wiener [1934, p. 84] were able to show that if

$$f \in L^2[-\pi, \pi] \quad \text{and} \quad \int_{-\pi}^{\pi} f(t) e^{\pm i\lambda_n t} \, dt = 0, \quad n = 1, 2, 3, \ldots,$$

then $f(t) = 0$ almost everywhere on $[-\pi, \pi]$, while Levinson [1935] derived the same conclusion when

$$f \in L^1[-\pi, \pi] \quad \text{and} \quad \int_{-\pi}^{\pi} f(t) e^{i\lambda_n t} \, dt = 0, \quad n = 1, 2, 3, \ldots.$$

(For an even stronger result, see Theorem 14.)

Problem 4. This is essentially Theorem I in Levinson [1940].

SECTION 2

The substance of this section is taken from Levinson [1940, Chap. I, Section 3].

Problem 3. Levinson [1940, p. 6]. Redheffer [1977, p. 34] has established the following sharp result for $p = 2$: If $\{\lambda_n\}$ is a sequence of real or complex numbers satisfying $|\lambda_n - n| \leqq L$ $(n = 0, \pm 1, \pm 2, \ldots)$, then the $L^2[-\pi, \pi]$ excess E of the system $\{e^{i\lambda_n t}\}$ satisfies

$$-\left(4L + \tfrac{1}{2}\right) < E \leqq 4L + \tfrac{1}{2}.$$

(The first result of this type is due to Paley and Wiener [1934, p. 94].)

Problem 4. Redheffer [1977, p. 45].

SECTION 3

Theorem 5. The first assertion of the theorem is implicit in the proof of Theorem XIX in Levinson [1940].

SECTION 4

Theorems 6, 7. Levinson [1936, 1940].

Theorem 8. See, e.g., Alexander and Redheffer [1967, Remark 3, p. 61].

Theorems 9, 10. Schwartz [1959, p. 102].

Problem 1. This result was proved by Paley and Wiener [1934, p. 89] under the added assumption that $\lim \lambda_n/n = 1$. That this condition must in fact hold whenever the system $\{e^{i\lambda_n t}\}$ is exact is a consequence of a well-known property of entire functions (see, e.g., Levinson [1940, p. 25]).

SECTION 5

Theorem 11. Redheffer [1957]. The theorem remains valid for complex sequences (Alexander and Redheffer [1967]; cf. Problem 1). When $p = 2$ and when the λ_n's are regularly distributed, then more can be said. Suppose that $\{\lambda_n\}$ is a real sequence such that the number of points of the sequence in the interval $(t, t+1)$ is uniformly bounded for $t \in (-\infty, \infty)$. Then each of the following conditions implies that the systems $\{e^{i\lambda_n t}\}$ and $\{e^{i\mu_n t}\}$ have the same excess in $L^2[-A, A]$:

1. $\sum |\lambda_n - \mu_n|^s < \infty$ for some positive number s;
2. $|\lambda_n - \mu_n| \leqq \varepsilon_n$, where ε_n decreases to zero and $\sum_{n=1}^{\infty} \varepsilon_n/n < \infty$.

(Sedleckiĭ [1977b]; it is also shown that the condition $\lambda_n - \mu_n \to 0$ is not sufficient to guarantee the equality of the excesses.) For further results in this direction, see Sedleckiĭ [1974, 1977b] and Redheffer [1977]. Although Theorem 11 is much weaker than Theorem 4 when $\mu_n = n$ $(n = 0, \pm 1, \pm 2, \ldots)$, the convergence criterion is sharp in the following sense: If $\{\varepsilon_n\}$ is an arbitrary sequence of positive numbers for which $\inf \varepsilon_n = 0$, then there exist real separated sequences $\{\lambda_n\}$ and $\{\mu_n\}$ such that

$$\sum |\lambda_n - \mu_n| \varepsilon_n < \infty \quad \text{but} \quad -\infty < E(\lambda) < E(\mu) < \infty$$

(Peterson [1974]). Here $E(\lambda)$ and $E(\mu)$ denote the respective excesses of $\{e^{i\lambda_n t}\}$ and $\{e^{i\mu_n t}\}$ in $L^2[-\pi, \pi]$.

Theorem 12. Elsner [1971]. The theorem was rediscovered by Young [1976a], and it is this proof that we have reproduced here. Sedleckiĭ [1978] has shown that in the spaces $L^1[-A, A]$ and $C[-A, A]$ the conclusion of the theorem is no longer true: Put $\lambda_n = n$ $(n = 0, \pm 1, \pm 2, \ldots)$ and $\mu_n = n$ for $n \leqq 0$, $\mu_n = n + i(-1)^n h$ for $n > 0$ (h is real and nonzero). Then the system $\{e^{i\mu_n t}\}$ has excess 1 in $L^1[-\pi, \pi]$, while in $C[-\pi, \pi]$ it is exact. Notice of course that the trigonometric system is exact in $L^1[-\pi, \pi]$ and has deficiency 1 in $C[-\pi, \pi]$. Whether or not the theorem remains valid in L^p, $1 < p < \infty$, is an open problem. It is essential that the two sequences $\{\lambda_n\}$ and $\{\mu_n\}$ lie in a horizontal strip. **Theorem (Sedleckiĭ [1985]).** *Let* $\varphi(x)$ *be an arbitrary* *nonnegative function defined on the nonnegative real axis and suppose that $\varphi(x)$ increases to infinity with x. Then there exist sequences $\{\lambda_n\}$ and $\{\mu_n\}$ of points lying in the region $\{z : |y| \leq \varphi(|x|)\}$ such that* $\operatorname{Re}\lambda_n = \operatorname{Re}\mu_n$ *while the excesses of $\{e^{i\lambda_n t}\}$ and $\{e^{i\mu_n t}\}$ in $L^2[-\pi, \pi]$ are unequal.*

Problem 1. Redheffer [1977, p. 11]. The references for Theorem 11 apply here as well.

SECTION 6

Theorem 14. Levinson [1935]. For a more elementary proof, see Redheffer [1977, p. 22].

Theorem 15. Schwartz [1959]. The theorem remains valid for complex λ_n's (Redheffer [1968, Remark 2, p. 105]).

Problem 9. Redheffer [1954, p. 60].

For an excellent account of the work of Beurling and Malliavin, see Kahane [1966] and Redheffer [1977]. (See also Koosis [1979].)

When a sequence $\{\lambda_n\}_{-\infty}^{\infty}$ of real or complex numbers is sufficiently regular, in the sense that $\sup_n |\lambda_n - n| < \infty$, the corresponding completeness radius is equal to π. The situation is dramatically different when completeness over a *union* of intervals is considered. **Theorem (Landau [1964]).** *Given $\varepsilon > 0$, there exists a real symmetric sequence $\{\lambda_n\}_{-\infty}^{\infty}$ with $|\lambda_n - n| < \varepsilon$ such that the system $\{e^{i\lambda_n t}\}$ is complete in C(S), where S is any finite union of intervals of the form $|t - 2n\pi| < \pi - \delta$, with n an integer and $\delta > 0$.*

CHAPTER 4

SECTION 1

Moment problems. The term *moment* was introduced into mathematical terminology by Stieltjes [1894]. In this classic paper the following problem, which he called the *problem of moments*, is stated and completely solved: Find a bounded nondecreasing function $\mu(t)$ on the interval $[0, \infty)$ such that its "moments" $\int_0^\infty t^n \, d\mu(t)$, $n = 0, 1, 2, \ldots$, have prescribed values

$$\int_0^\infty t^n \, d\mu(t) = c_n, \quad n = 0, 1, 2, \ldots.$$

Subsequent variants of the problem include the *Hamburger moment problem* (Hamburger [1920]), in which the interval $[0, \infty)$ is replaced by $(-\infty, \infty)$; the

Hausdorff moment problem (Hausdorff [1923]), in which $[0, \infty)$ is replaced by $[0, 1]$; and the *trigonometric moment problem*

$$\int_0^{2\pi} e^{int}\, d\mu(t) = c_n, \quad n = 0, \pm 1, \pm 2, \ldots,$$

considered by Akhiezer and Krein [1934]. For a detailed study of these and other moment problems, see Shohat and Tamarkin [1943] and Akhiezer [1965]. A different approach, based on the spectral theorem for self-adjoint operators, is found in Dunford and Schwartz [1963, pp. 1250–1256]. Other references: Akhiezer and Krein [1962], Natanson [1965, pp. 126–170], and Krein and Nudelman [1977].

Theorem 1. Paley and Wiener [1934, p. 89], under the additional assumption that $\lim \lambda_n / n = 1$. The proof in the text is from Young [1979].

Example 3. This example appears, in essence, in Paley and Wiener [1934, pp. 114–116]. For additional examples and applications of Lagrange type series, especially to uniqueness and growth theorems, see Boas [1954, Chaps. 9, 10] and Levin [1964, Chap. IV]; see also Section 5.

Theorem 2. Hahn [1927, p. 216] proves a slightly more general result when H is a Banach space. For this and related results, see Dunford and Schwartz [1958, pp. 86–87].

SECTION 2

Bessel sequences and *Riesz–Fischer sequences* were introduced and extensively studied by Bari [1951]. Her terminology is slightly different from ours.

Theorem 3. The original form of this theorem, with $H = L^2[0, 1]$, is due to Boas [1941]. It was rediscovered by Bari [1951], where a number of other equivalent conditions are also given.

Interpolation problems in spaces of analytic functions. See Carleson [1958], Shapiro and Shields [1961], Hoffman [1962, pp. 194–208], Bochner [1964], Landau [1967], Rosenbaum [1967, 1968], Levin [1969], Duren [1970, Chap. 9], Earl [1970], Sedleckiĭ [1971b, 1973], Binmore [1972], Duren and Williams [1972], Taylor and Williams [1972], Levin and Lyubarskiĭ [1974, 1975], Švedenko [1974], Young [1974a, 1976b], Švedenko and Turku [1975], Neville [1977], Hruščev, Nikolskiĭ, and Pavlov [1981], Beurling [1989], Berndtsson and Ortega-Cerdà [1995], Seip [1995, 1998], Hedenmalm, Richter, and

Seip [1996], Lyubarskiĭ and Seip [1997], Young [1997, 1998], Schuster and Seip [1998], and Ortega-Cerdà and Seip [1999].

Problem 5. The suggested proof is due to Fejér and Riesz [1921]. The inequality named after them is valid whether the coefficients c_n are real or complex, and has important function-theoretic applications (see, e.g., Duren [1970]). For additional proofs of Hilbert's inequality, together with applications, generalizations, and historical remarks, see Hardy, Littlewood, and Pólya [1952, Chap. IX].

Problem 8. Beurling [1989, pp. 341–365] gave a complete solution to the interpolation problem in the Bernstein space $\boldsymbol{B}_\tau(\tau > 0)$: A real sequence $\Lambda = \{\lambda_n\}$ is an interpolating sequence for $\boldsymbol{B}_\tau$ if and only if Λ is separated and has *upper uniform density* less than τ/π. If Λ is a separated sequence of real numbers, we denote by $n^+(r)$ the maximum number of terms of Λ to be found in an interval of length r. Then the limit, $\lim_{r\to\infty} n^+(r)/r$, which always exists, is called the "upper uniform density" of Λ. Beurling's work was presented in a series of seminar lectures given in 1959-60 at the Institute for Advanced Study in Princeton but it was not published until 1989 when his collected works appeared. An extension to complex sequences was given by Ortega-Cerdà and Seip [1999]. For related results, see the *Notes* for Theorem 4.5.

SECTION 3

Theorem 4. Titchmarsh [1928]. Many proofs have been given; see, e.g., Paley and Wiener [1934, p. 117], Ingham [1936], and Amerio [1941]. The following more general result, which extends the Hausdorff–Young theorem, appeared without proof in Titchmarsh [1925] and was subsequently proved in Titchmarsh [1965]: If $\{\lambda_n\}$ is a separated sequence of real numbers, if $p \geqq 2$, and if $\{c_n\} \in l^{p/(p-1)}$, then the series $\sum c_n e^{i\lambda_n t}$ converges in the mean of order p over every finite interval. The λ_n's may be complex provided they are separated and $\sup_n |\operatorname{Im} \lambda_n| < \infty$ (see, e.g., Young [1974b]). For a discussion of various other types of convergence of nonharmonic Fourier series, see Kac [1941], Bellman [1943], and Cossar [1975/76].

Theorem 5. Both the theorem and the counterexample following it are due to Ingham [1936]. A similar but less precise result had already been obtained by Paley and Wiener [1934, p. 118]. The following impressive generalization is due to Beurling [1989, pp. 341–365]. If $\{\lambda_n\}$ is a separated sequence of real numbers with upper uniform density less than A/π, then the system $\{e^{i\lambda_n t}\}$ is a Riesz-Fischer sequence in $L^2[-A, A]$. The definition of "upper uniform density" is given in these *Notes* (see Section 2, Problem 8). In the opposite direction, Landau [1967] had already shown that if $\{e^{i\lambda_n t}\}$ is a Riesz-Fischer sequence in $L^2[-A, A]$, then $\{\lambda_n\}$ is separated and has upper uniform density less than or equal to A/π. These results have been used by Jaffard [1991], Seip [1995], and Young [1997] to investigate frames and Riesz bases of complex exponentials in $L^2[-A, A]$. For a different extension of Theorem 5, see Jaffard, Tucsnak, and Zuazua [1997].

Problem 3. Ingham [1936].

Problem 4. Young [1976c, p. 317].

SECTION 4

Theorem 7. Gurevich [1941, Theorem 3], under additional hypotheses. The proof in the text is from Singer [1970, Part I, Theorem 8.1d].

Theorem 8. Put another way, this is simply the assertion that a system of vectors in a separable Hilbert space is a Riesz basis if and only if it possesses a biorthogonal system which is also a Riesz basis. The formulation in the text appears explicitly in Korobeĭnik [1976].

Problem 3. Grinblium [1948]. See, e.g., Singer [1970, p. 75].

SECTION 5

Interpolation in the Paley–Wiener space was studied extensively by Young [1974a]. For additional facts about functions of sine type, see Levin [1961, 1969, 1996], Sedleckiĭ [1970, 1972a], Levin and Ostrovskiĭ [1980], and Avdonin and Joó [1989].

Theorem 10. Levin [1961] was the first to prove that the system is a basis for $L^2[-\pi, \pi]$. Golovin [1964] showed that it is in fact a Riesz basis. The proof given in the text is due essentially to Levin [1969]. It is easy to exhibit Riesz bases $\{e^{i\lambda_n t}\}$ for $L^2[-\pi, \pi]$ that do not arise in this way. If $0 < \varepsilon < \frac{1}{4}$ and

$$\lambda_n = \begin{cases} n + \varepsilon & n > 0 \\ 0 & n = 0 \\ n - \varepsilon & n < 0 \end{cases}$$

then Kadec's $\frac{1}{4}$-Theorem applies but $\{\lambda_n\}$ is not the set of zeros of a function of sine type (Redheffer and Young [1983]). For an impressive generalization of Levin's theorem, see Katsnelson [1971] (see also Avdonin [1977]). A generalization in another direction has been given by Levin and Lyubarskiĭ [1974].

The duality between basis theory and interpolation theory has been exploited in many different directions. The following papers deserve mention: Binmore [1972], Dragilev, Zaharjuta, and Korobeĭnik [1974], Levin and Lyubarskiĭ [1975], Korobeĭnik [1976, 1979], Oskolkov [1978], Martirosjan [1979], Young [1987], Higgins [1994], Seip [1995], and Lyubarskiĭ and Seip [1997].

Problem 3. Sedleckiĭ [1970].

Problem 7. This result is due to Levin [1969], who shows that the mapping $l^p \to \boldsymbol{E}_\pi^p$ thus defined is *onto* $\boldsymbol{E}_\pi^p$. Lyubarskiĭ and Seip [1997] have characterized the complete interpolating sequences for the spaces $\boldsymbol{E}_\pi^p (1 < p < \infty)$ in terms of the "Muckenhoupt condition". When $p = 2$ their description coincides with Pavlov's characterization of the class of exponential Riesz bases for $L^2[-\pi, \pi]$ (see the *Notes* for the exponential Riesz basis problem in Section 8). Interpolating sequences for $\boldsymbol{E}_\pi^p$ are further investigated in Young [1997]: **Theorem**. *If* $\{\lambda_n\}$ *is a sequence of real numbers with upper uniform density less than* 1 *and if* $\{\mu_n\}$ *is an arbitrary bounded sequence of reals, then* $\{\lambda_n + i\mu_n\}$ *is an interpolating sequence for* $\boldsymbol{E}_\pi^p$ *whenever* $1 \leq p \leq \infty$. This generalizes a classical theorem due to Ingham (see Theorem 4.5). For an informative discussion of the discrete nature of the spaces $\boldsymbol{E}_\pi^p$, see Eoff [1995].

SECTION 6

Proposition 3 and Corollary 1. Young [1974a].

Lemma 3. Duffin and Schaeffer [1952].

Lemma 4. This is a special case of a theorem of Bade and Curtis [1966, p. 393]. The elementary proof given in the text is from Stray [1972].

Theorem 11. Young [1976b].

Problem 3. Ullrich [1989], Jaffard [1991].

Open problem. Let $\{\lambda_n\}$ and $\{\mu_n\}$ be two sequences of points lying in a fixed horizontal strip and suppose that $\operatorname{Re}\lambda_n = \operatorname{Re}\mu_n$ for all n. If $\{\lambda_n\}$ is an interpolating sequence for the Paley–Wiener space $\boldsymbol{P}$, does it follow that $\{\mu_n\}$ is also an interpolating sequence for $\boldsymbol{P}$?

SECTION 7

General references. The first systematic investigation of frames in Hilbert space was made by Duffin and Schaeffer [1952]. Since then, and especially over the past two decades, research in this area has flourished. Frames now play a fundamental role not only in abstract mathematics but in such applied areas as signal and image processing, data compression, sampling theory, and wavelet analysis. P. G. Casazza has prepared two general surveys on the theory and applications of frames: "The art of frame theory" (Casazza [2000]), focusing on abstract frames, and "Modern tools for Weyl-Heisenberg (Gabor) frame theory" (to appear in *Advances in Imaging and Electron Physics*), focusing on applications. For a wide range of applications to signal and image processing, data compression, and sampling theory, see Benedetto and Frazier [1994]. For applications to wavelets, see Daubechies, Grossmann, and Meyer [1986], Heil and Walnut [1989], Daubechies [1992], Meyer [1992], and Mallat [1999].

Stability of frames. The classical theorem of Paley and Wiener on the stability of orthonormal bases in Hilbert space (Theorem 1.13) has been generalized in many ways and in many different directions. Christensen [1995a] has established the following impressive generalization for frames.

Theorem. *Let $\{f_1, f_2, f_3, \ldots\}$ be a frame in a separable Hilbert space H, with frame bounds A and B, and suppose that λ and μ are nonnegative constants such that $\lambda + \mu/\sqrt{A} < 1$. If $\{g_1, g_2, g_3, \ldots\}$ is a sequence of elements of H such that*

$$\left\|\sum c_n(f_n - g_n)\right\| \le \lambda \left\|\sum c_n f_n\right\| + \mu\sqrt{\sum |c_n|^2}$$

for every finite sequence $\{c_n\}$ of scalars, then $\{g_1, g_2, g_3, \ldots\}$ is a frame with bounds $A(1 - (\lambda + \mu/\sqrt{A}))^2$ and $B(1 + \lambda + \mu/\sqrt{B})^2$.

For other stability results, see Balan [1997], Christensen [1995b, 1996b, 1997], Christensen and Heil [1997], Favier and Zalik [1995], Jing [1999], Su and Zhou [1999], Sun and Zhou [1999], and Young [1997].

Frames containing a Riesz basis. A frame need not contain a Riesz basis. In fact, there exists a frame of complex exponentials $\{e^{i\lambda_n t}\}$ in $L^2[-\pi, \pi]$ that cannot be made into a Riesz basis by removing from the set a suitable number of exponentials (Seip [1995] gives the counterexample $\lambda_n = n(1 - |n|^{-1/2})$ for $|n| > 1$). Casazza and Christensen [1996] have shown that, if $\{f_1, f_2, f_3, \ldots\}$ is a sequence of vectors in a separable Hilbert space H with the property that every subset is a frame in its closed linear span, then $\{f_1, f_2, f_3, \ldots\}$ contains a Riesz basis. (See also Christensen [1996a] and Casazza and Christensen [1998a].) Casazza [1999] has proved that there exists a normalized frame that contains a Schauder basis but not a Riesz basis. A frame $\{f_1, f_2, f_3, \ldots\}$ is said to be a *near-Riesz basis* for H if the removal of some *finite* subset of its elements leaves a Riesz basis. **Theorem (Holub [1994]).** *A frame for a Hilbert space is a near-Riesz basis if and only if it is "Besselian".* (A frame $\{f_1, f_2, f_3, \ldots\}$ is said to be *Besselian* provided the series $\Sigma c_n f_n$ converges if and only if the series $\Sigma |c_n|^2$ converges.) Casazza and Christensen [1998b] have proved that a frame need not even contain a Schauder basis.

Gabor frames. In his seminal paper of 1946, Gabor considered the system of functions

$$g_{mn}(x) = e^{2\pi imx} g(x-n) \qquad m, n \in \mathbf{Z}$$

generated by translations and modulations of a single function g in $L^2(\boldsymbol{R})$, and he proposed their use in communication theory (Gabor [1946]). Two well-known choices for g for which the system $\{g_{mn}\}_{m,n\in\mathbf{Z}}$ constitutes an orthonormal basis for $L^2(\boldsymbol{R})$ are

$$g(x) = \begin{cases} 1 & \text{if } 0 \le x \le 1 \\ 0 & \text{otherwise} \end{cases}$$

and

$$g(x) = \frac{\sin \pi x}{\pi x}.$$

If the system $\{g_{mn}\}_{m,n\in\mathbf{Z}}$ is a frame in $L^2(\boldsymbol{R})$, then it is called a **Gabor frame** (the term **Weyl-Heisenberg frame** is also used). The famous *Balian-Low theorem* asserts that for every frame of this type, either the generator g or its Fourier transform $\hat{g}$ must be badly localized:

$$\int_{-\infty}^{\infty} x^2 |g(x)|^2 \, dx = \infty \quad \text{or} \quad \int_{-\infty}^{\infty} y^2 |\hat{g}(y)|^2 \, dy = \infty.$$

The result was originally stated for orthonormal bases, independently by Balian [1981] and Low [1985]. Subsequently, Battle [1988] discovered an elegant proof based on the uncertainty principle. This proof was later generalized to frames by Daubechies and Janssen [1993]. For a survey of the Balian-Low theorem see Benedetto, Heil, and Walnut [1995].

To avoid the restrictions imposed by the Balian-Low theorem, it is natural to consider systems of the form

$$g_{mn}(x) = e^{imax} g(x-nb) \qquad m, n \in \mathbf{Z}$$

where a and b are fixed positive numbers and g belongs to $L^2(\boldsymbol{R})$. If $ab > 2\pi$ then the system can never be a frame; if $ab = 2\pi$ then such frames exist, but they have bad space-frequency localization; if $ab < 2\pi$ then frames with excellent space-frequency localization are possible (see, e.g., Daubechies [1992]).

Wavelet frames. Wavelets made their appearance in the mid 1980s as a synthesis of ideas being explored independently in mathematics, theoretical physics, and engineering. Researchers working in diverse areas were hoping to find efficient algorithms for decomposing functions, or signals, into elementary functions that would combine the best features of the trigonometric and Haar systems. Wavelet analysis replaces the complex exponential building blocks with more flexible units — orthonormal bases generated by *translations and dilations* of a single function. By definition, an **orthonormal wavelet basis** is an orthonormal basis for $L^2(\boldsymbol{R})$ consisting of functions of the form

$$\Psi_{mn}(x) = 2^{m/2}\Psi(2^m x - n) \qquad m, n \in \boldsymbol{Z}$$

where Ψ is a unit vector in $L^2(\boldsymbol{R})$. The factor $2^{m/2}$ is included to ensure that each basis element has unit length. The function Ψ that generates the basis is called a **wavelet**. Two well-known choices for Ψ for which the system $\{\Psi_{mn}\}_{m,n\in\boldsymbol{Z}}$ constitutes an orthonormal basis for $L^2(\boldsymbol{R})$ are

$$\Psi(x) = \begin{cases} 1, & 0 \le x < 1/2, \\ -1, & 1/2 \le x < 1, \\ 0, & \text{otherwise} \end{cases}$$

(the **Haar wavelet**) and

$$\Psi(x) = \frac{\sin 2\pi x - \sin \pi x}{\pi x}$$

(the **Shannon wavelet**). If the system $\{\Psi_{mn}\}_{m,n\in\boldsymbol{Z}}$ is a frame in $L^2(\boldsymbol{R})$, then it is called a **wavelet frame**. While a C^∞ function with exponential decay can never generate an orthonormal wavelet basis, it can generate a wavelet frame. An example is the "Mexican hat function", the second derivative of the Gaussian $e^{-x^2/2}$; if we normalize it so that $\|\Psi\| = 1$ and $\Psi(0) > 0$, then

$$\Psi(x) = \frac{2}{\sqrt{3}}\pi^{-1/4}(1 - x^2)e^{-x^2/2}.$$

(See Daubechies [1992], who has reported frame bounds of 3.223 and 3.596, so that the frame is close to tight.) This example is historically important because of its use by Grossmann and Morlet in the numerical analysis of seismic signals (Goupillaud, Grossmann, and Morlet [1984/85]). In the past fifteen years, wavelet theory has succeeded in revolutionizing modern harmonic analysis. Wavelets and their generalizations provide unconditional bases for most of the classical function spaces; they offer flexible methods for studying integral operators and partial differential equations; they are particularly effective for analyzing functions having a multifractal structure; and, they remain an indispensable tool for signal and image processing. (See the "general references" at the beginning of this section.)

Nonharmonic Fourier frames. Nearly 50 years after Duffin and Schaeffer introduced the theory of frames, J. Ortega-Cerdà and K. Seip have solved the problem of characterizing those real sequences that generate nonharmonic Fourier frames in $L^2[-\pi, \pi]$. Their work ("On Fourier frames," preprint) is an elegant interplay of three themes: de Branges' theory of Hilbert spaces of entire functions, Pavlov's characterization of exponential Riesz bases (see the *Notes* for Chapter 4, Section 8), and the approximation of subharmonic functions.

Problem 4. Christensen [1996b].

Problem 5. Christensen [1993].

Aldroubi [1995] has given a complete characterization of all the frames of a given Hilbert space. For additional results on frames, see Balan [1999], Benedetto [1992, 1994, 1998, 1999], Benedetto and Li [1998], Benedetto and Walnut [1994], Casazza [1997, 1998], Casazza and Christensen [2000], Casazza, Christensen, and Janssen [1999], Casazza, Han, and Larson [1999], Christensen [2000], Christensen, Deng, and Heil [1999], Chui and He [2000], Chui and Shi [1993a, 1993b], Daubechies and Grossmann [1988], Favier and Zalik [1997], Feichtinger [1990], Gabardo [1993, 1995], Gröchenig [1991], Gröchenig and Ron [1998], Heil [1990], Heller [1992], Jaffard [1991], Janssen [1995, 1996], Kaiser [1994], Lawton [1990], Li [1995], Lindner [1999], Lyubarskiĭ [1992], Ron and Shen [1995], Seip and Ulanovskiĭ [1996], Tolimiera and Orr [1995], Volkmer [1995], and Voss [1999].

SECTION 8

Theorem 13. Duffin and Schaeffer [1952]. Using Christensen's "Paley–Wiener theorem for frames" (see the *Notes* for Chapter 4, Section 7), Balan [1997] and Christensen [1997] have independently established the following generalization of Kadec's $\frac{1}{4}$-Theorem: *Let* $\{\lambda_n\}$ *and* $\{\mu_n\}$ *be two sequences of real numbers such that for all* n,

$$|\lambda_n - \mu_n| \leq L < \frac{1}{4} - \frac{1}{\pi} \arcsin\left(\frac{1}{\sqrt{2}}\left(1 - \sqrt{\frac{A}{B}}\right)\right);$$

here A *and* B *are positive constants with* $A \leq B$. *If the system* $\{e^{i\lambda_n t}\}$ *is a frame in* $L^2[-\pi, \pi]$, *with bounds* A *and* B, *then the system* $\{e^{i\mu_n t}\}$ *is also a frame.* A sharper estimate is given by Su and Zhou [1999].

Corollary. Young [1975b]. The corollary remains valid if the term "Riesz basis" is replaced by "unconditional basis" in both the hypothesis and the conclusion. This was established by Hruščev, Nikolskiĭ, and Pavlov [1981, p. 295] using deep function-theoretic results. A completely elementary proof is given in Young [1987].

Open problem. Does the corollary to Theorem 13 remain valid if the term "Riesz basis" is replaced by "Schauder basis" in both the hypothesis and the conclusion?

Theorem 14. Duffin and Schaeffer [1952].

Corollary 1. Young [1974a, p. 108].

Corollary 2. This result is stated, with an incorrect proof, in Kadec [1964], in which a theorem of Duffin and Schaeffer is misquoted. Katsnelson [1971] has proved an even stronger result, which contains both Corollary 2 and Theorem 10 as special cases (see also Avdonin [1977]). The proof in the text is due to Young [1975a]. Recently, Pavlov [1979] has obtained a striking criterion —based on the "Muckenhoupt condition" —for a system of complex exponentials to form a Riesz basis for $L^2[-\pi, \pi]$.

The exponential Riesz basis problem. The main result in the theory of Riesz bases of complex exponentials $\{e^{i\lambda_n t}\}$ in $L^2[-\pi, \pi]$ is Pavlov's characterization of such bases in terms of the entire generating function

$$F(z) = \lim_{R\to\infty} \prod_{|\lambda_n|<R} \left(1 - \frac{z}{\lambda_n}\right).$$

Since multiplication by e^{ax} is a bounded invertible operator on $L^2[-\pi, \pi]$, we may assume that the λ_n lie in a horizontal strip of finite width *above the real axis*. Pavlov's necessary and sufficient condition for basicity is that the λ_n be separated, that F be an entire function of exponential type π, and that the function $w(x) = |F(x)|^2$ satisfy the "Muckenhoupt condition"

$$\sup_I \left(\frac{1}{|I|}\int_I w(x)\,dx \cdot \frac{1}{|I|}\int_I \frac{1}{w(x)}\,dx\right) < \infty;$$

here the supremum is taken over all bounded intervals I on the real line and $|I|$ denotes the length of I (Pavlov [1979]). Using Pavlov's criterion, Hruščev [1979] has derived both Corollary 2 and Theorem 10—and more. For an excellent account of the exponential Riesz basis problem and its applications to problems of control, see Avdonin and Ivanov [1995].

Unconditional bases of complex exponentials. There exist unconditional bases $\{e^{i\lambda_n t}\}$ for $L^2[-\pi, \pi]$ for which the λ_n do not lie in a horizontal strip of finite width. Of course, such bases cannot be Riesz bases. The following striking theorem is due to Vasyunin (see Hruščev, Nikolskiĭ, and Pavlov [1981, p. 301]).

Theorem. *Let $\{\lambda_n\}$ be a set of points in the upper half-plane,* $\operatorname{Im} z > 0$, *satisfying the "Carleson condition"*

$$\inf_n \prod_{k \neq n} \left| \frac{\lambda_k - \lambda_n}{\lambda_k - \overline{\lambda}_n} \right| > 0,$$

and such that $\lim_{n\to\infty} \operatorname{Im} \lambda_n = \infty$. *Then for each positive number A, there is a set $\{\mu_n\}$ containing $\{\lambda_n\}$ such that $\{e^{i\mu_n t}\}$ is an unconditional basis for* $L^2[-A, A]$.

Carleson formulated the condition that bears his name in his solution to the classical interpolation problem for bounded analytic functions in the upper half-plane, $\operatorname{Im} z > 0$ (Carleson [1958]; see also Shapiro and Shields [1961]). The Carleson condition also characterizes those families of exponentials $\{e^{i\lambda_n t}\}$ that form unconditional bases for their closed linear spans in $L^2(0, \infty)$. Using this result, Hruščev, Nikolskiĭ, and Pavlov [1981] have given a complete characterization of those sequences $\{\lambda_n\}$ of points lying in the upper half-plane for which the family of exponentials $\{e^{i\lambda_n t}\}$ forms an unconditional basis for $L^2[-\pi, \pi]$.

Problem 2. Young [1975b]. A stronger result appears in Young [1976c].

Open problem. If the system of complex exponentials $\{e^{i\lambda_n t}\}$ is an unconditional basis for its closed linear span in $L^2[-\pi, \pi]$, can it be enlarged to an unconditional basis $\{e^{i\mu_n t}\}$ for $L^2[-\pi, \pi]$?

If we replace the term "unconditional basis" with "Riesz basis" throughout, then the problem has a negative solution: the counterexample

$$\lambda_n = n(1 + |n|^{-1/2}) \quad \text{for } |n| > 1$$

was discovered by Seip [1995].

SECTION 9

Theorem 15. Duffin and Schaeffer [1952]. The first result of this sort is due to Paley and Wiener [1934, p. 113], under the assumption that each λ_n is real and $\sup_n |\lambda_n - n| < 1/\pi^2$. (Cf. Walsh [1921], where it is shown that if λ_n is "very near" to n, then each function in $L^2[0, \pi]$ has both an ordinary Fourier cosine series $\sum_{n=0}^{\infty} a_n \cos nt$ and a nonharmonic Fourier cosine series $\sum_{n=0}^{\infty} b_n \cos \lambda_n t$, which are uniformly equiconvergent on $[0, \pi]$.) The following impressive generalization for L^p is due to Sedleckiĭ [1971a].

Theorem. *If* $\{\lambda_n\}$ *is a sequence of real or complex numbers such that*

$$|\operatorname{Re}\lambda_n - n| \leqq L < \frac{p-1}{2p}, \quad \sup_n |\operatorname{Im}\lambda_n| < \infty \quad (n = 0, \pm 1, \pm 2, \dots)$$

for some p, $1 < p \leqq 2$, *then the set* $\{e^{i\lambda_n t}\}$ *is complete in* $L^q[-\pi, \pi]$ *and possesses a unique biorthogonal set* $\{h_n(t)\}$ *in* $L^q[-\pi, \pi]$ *such that for every* $f \in L^p[-\pi, \pi]$ *the series*

$$\sum_{-\infty}^{\infty} \left(e^{int} \frac{1}{2\pi} \int_{-\pi}^{\pi} f(x) e^{-inx}\, dx - e^{i\lambda_n t} \int_{-\pi}^{\pi} f(x) h_n(x)\, dx \right)$$

converges to zero uniformly on every compact subset of $(-\pi, \pi)$.

The theorem above generalizes a classical result of Levinson [1940, Chap. IV], in which the λ_n's are assumed to satisfy the more restrictive condition $|\lambda_n - n| \leqq L < (p-1)/2p$ $(n = 0, \pm 1, \pm 2, \dots)$. Levinson showed that his result is sharp in the sense that if $L = (p-1)/2p$, then the conclusions of the theorem no longer hold. For a variety of other equiconvergence results, in which the λ_n's are the zeros of an entire function belonging to a special class (for example, the class of functions of sine type), see Sedleckiĭ [1970, 1972a, 1972b, 1975a, 1975b, 1977a] (see also Verblunsky [1956]).

Problem 4. Duffin and Schaeffer [1952].

REFERENCES

Ahlfors, L. V. [1979]. "Complex Analysis," 3rd ed., McGraw-Hill, New York.

Akhiezer, N. I. [1956]. "Theory of Approximation." Ungar, New York.

Akhiezer, N. I. [1965]. "The Classical Moment Problem." Hafner, New York.

Akhiezer, N. I., and Krein, M. G. [1934]. Ueber Fouriersche Reihen beschränkter summierbarer Funktionen und ein neues Extremumproblem. I, II. *Commun. Soc. Math. Kharkoff* **9**, 9–28; **10**, 3–32.

Akhiezer, N. I., and Krein, M. G. [1962]. "Some Questions in the Theory of Moments," Translations of Mathematical Monographs, Vol. 2., Am. Math. Soc., Providence, Rhode Island.

Aldroubi, A. [1995]. Portraits of frames. *Proc. Amer. Math. Soc.* **123**, 1661–1668.

Alexander, W. O., Jr., and Redheffer, R. M. [1967]. The excess of sets of complex exponentials. *Duke Math. J.* **34**, 59–72.

Al′tman, M. [1949]. On bases in Hilbert space. *Doklady Akad. Nauk SSSR* **69**, 483–485.

Amerio, L. [1941]. Sulla convergenza in media della serie $\sum_0^\infty a_n e^{i\lambda_n x}$. *Ann. Sc. Norm. Super. Pisa* **10**, 191–198.

Anderson, J. M. [1978]. Müntz–Szász theorems and lacunary entire functions. *Linear Spaces Approx., Proc. Conf., Oberwolfach, 1977. Int. Ser. Numer. Math.* **40**, 491–501.

Aronszajn, N. [1950]. Theory of reproducing kernels. *Trans. Am. Math. Soc.* **68**, 337–404.

Arsove, M. G. [1958]. Similar bases and isomorphisms in Fréchet spaces. *Math. Ann.* **135**, 283–293.

Arutjunjan, F. G. [1972]. Bases of the spaces $L_1[0, 1]$ and $C[0, 1]$. *Math. Notes* **11**, 152–157.

Avdonin, S. A. [1977]. On the question of Riesz bases consisting of exponential functions in L^2. *J. Soviet Math.* **8**, 130–131.

Avdonin, S. A. [1979]. On Riesz bases of exponentials in L^2. *Vestnik Leningrad Univ.* **7**, 203–211.

Avdonin, S. A. [1991]. The existence of basis subfamilies of a Riesz basis from exponentials. *Vestnik Leningrad Univ. Math.* **24**, 59–60.

Avdonin, S. A., Horvath, M., and Joó, I. [1989]. Riesz bases from elements of the form $x^k e^{i\lambda_k x}$. *Vestnik Leningrad Univ. Math.* **22**, 1–6.

Avdonin, S. A., and Ivanov, S. A. [1984]. Series bases of exponentials and the problem of the complete damping of a system of strings. *Soviet Physics Dokl.* **29**, 182–184.

Avdonin, S. A., and Ivanov, S. A. [1995]. "Families of Exponentials." Cambridge University Press, Cambridge.

Avdonin, S. A., and Joó, I. [1989]. Riesz bases of exponentials and sine type functions. *Acta Math. Hung.* **51**, 3–14.

Avdonin, S. A., Ivanov, S. A., and Io, I. [1990]. Families of exponentials and controllability of a rectangular membrane. *Studia Sci. Math. Hungar.* **25**, 291–306.

Avdonin, S. A., Ivanov, S. A., and Joó, I. [1995]. Exponential series in the problems of initial and pointwise control of a rectangular vibrating membrane. *Studia Sci. Math. Hungar.* **30**, 243–259.

Avdonin, S. A., Ivanov, S. A., and Russell, D. L. [1999]. Exponential bases in Sobolev spaces in control and observation problems. *Optimal control of partial differential equations (Chemnitz, 1998)*, 33–42, *Internat. Ser. Numer. Math.* **133**, Birkhäuser, Basel.

Babenko, K. I. [1948]. On conjugate functions. *Dokl. Akad. Nauk SSSR* **62**, 157–160.

Bade, W. G., and Curtis, P. C., Jr. [1966]. Embedding theorems for commutative Banach algebras. *Pac. J. Math.* **18**, 391–409.

Balan, R. [1997]. Stability theorems for Fourier frames and wavelet Riesz bases. *J. Fourier Anal. Appl.* **3**, 499–504.

Balan, R. [1999]. Equivalence relations and distances between Hilbert frames, *Proc. Amer. Math. Soc.* **127**, 2353–2366.

Balian, R. [1981]. Un principe d'incertitude fort en théorie du signal ou en mécanique quantique. *C.R. Acad. Sci. Paris*, Sér. II, **292**, 1357–1361.

Banach, S. [1932]. "Théorie des opérations linéaires." Warsaw. Reprinted by Chelsea, Bronx, New York, 1955.

Bari, N. K. [1951]. Biorthogonal systems and bases in Hilbert space. *Učen Zap. Mosk. Gos. Univ. 148, Mat.* **4**, 69–107.

Battle, G. [1988]. Heisenberg proof of the Balian-Low theorem. *Letters in Math. Physics* **15**, 175–177.

Bellman, R. [1943]. Convergence of non-harmonic Fourier series. *Duke Math. J.* **10**, 551–552.

Benedetto, J. J. [1992]. Irregular sampling and frames. *Wavelets*, 445–507, *Wavelet Anal. Appl.* **2**, Academic Press, Boston.

Benedetto, J. J. [1994]. Frame decompositions, sampling, and uncertainty principle inequalities. *In* "Wavelets: Mathematics and Applications," Benedetto and Frazier, editors, pp. 247–304. CRC Press, Boca Raton.

Benedetto, J. J. [1997]. "Harmonic Analysis and Applications." CRC Press, Boca Raton.

Benedetto, J. J. [1998]. Noise reduction in terms of the theory of frames. *Signal and image representation in combined spaces*, 259–284, *Wavelet Anal. Appl.* **7**, Academic Press, San Diego.

Benedetto, J. J. [1999]. Frames, sampling, and seizure prediction. *Advances in Wavelets (Hong Kong, 1997)*, 1–25, Springer, Singapore.

Benedetto, J. J., and Frazier, M. W. [1994]. "Wavelets: Mathematics and Applications." CRC Press, Boca Raton.

Benedetto, J. J., and Li, S. [1998]. The theory of multiresolution analysis frames and applications to filter banks. *Appl. Comput. Harmon. Anal.* **5**, 389–427.

Benedetto, J. J., and Walnut, D. F. [1994]. Gabor frames for L^2 and related spaces. *In* "Wavelets: Mathematics and Applications," Benedetto and Frazier, editors, pp. 97–162. CRC Press, Boca Raton.

Benedetto, J. J., Heil, C., and Walnut, D. F. [1995]. Differentiation and the Balian-Low theorem. *J. Fourier Anal. Appl.* **1**, 355–402.

Berberian, S. K. [1961]. "Introduction to Hilbert Space." Oxford Univ. Press, London and New York.

Bergman, S. [1947]. "Sur les fonctions orthogonales de plusieurs variables complexes avec les applications à la théorie des fonctions analytiques." Gauthier-Villars, Paris.

Bergman, S. [1950]. "The Kernel Function and Conformal Mapping." Am. Math. Soc., New York.

Berndtsson, B., and Ortega-Cerdà, J. [1995]. On interpolation and sampling in Hilbert spaces of analytic functions. *J. Reine Angew. Math.* **464**, 109–128.

Bernstein, S. N. [1912a]. Démonstration du théorème de Weierstrass fondée sur le calcul de probabilité. *Soobšč. Har'kov. Mat. Obšč. (2)* **13**, 1–2.

Bernstein, S. N. [1912b]. Sur l'ordre de la meilleure approximation des fonctions continues par des polynômes de degré donné. Doctoral Dissertation (French translation). *Mem. de l'Académie Royale de Belgique.* Cl. Sci., Collection (2) **4**, 1–103.

Bernstein, S. N. [1923]. Sur une propriété des fonctions entières. *C. R. Acad. Sci.* **176**, 1603–1605.

Besicovitch, A. S. [1932]. "Almost Periodic Functions." Cambridge Univ. Press, London.

Bessaga, C., and Pelczynski, A. [1958]. On bases and unconditional convergence of series in Banach spaces. *Stud. Math.* **17**, 151–164.

Beurling, A. [1989]. "The Collected Works of Arne Beurling, Volume 2, Harmonic Analysis" (L. Carleson, P. Malliavin, J. Neuberger and J. Wermer, editors), pp. 341–365, Birkhäuser, Boston.

Beurling, A., and Malliavin, P. [1967]. On the closure of characters and the zeros of entire functions. *Acta Math.* **118**, 79–93.

Billard, P. [1972]. Bases dans H et bases de sous-espaces de dimension finie dans A. *Linear Operators Approx., Proc. Conf., Oberwolfach, 1971. Int. Ser. Numer. Math.* **20**, 310–324.

Binmore, K. G. [1972]. Interpolation, approximation, and gap series. *Proc. London Math. Soc.* **25**, 751–768.

Birkhoff, G. [1935]. Orthogonality in linear metric spaces. *Duke Math. J.* **1**, 169–172.

Birkhoff, G., and Rota, G. C. [1960]. On the completeness of Sturm–Liouville expansions. *Am. Math. Mon.* **67**, 835–841.

Birkhoff, G., and Rota, G. C. [1962]. "Ordinary Differential Equations." Ginn, New York.

Boas, R. P., Jr. [1937]. The derivative of a trigonometric integral. *J. London Math. Soc.* **12**, 164–165.

Boas, R. P., Jr. [1938]. Representations for entire functions of exponential type. *Ann. Math.* **39**, 269–286 (correction, **40**, 948).

Boas, R. P., Jr. [1940]. Expansions of analytic functions. *Trans. Am. Math. Soc.* **48**, 467–487.

Boas, R. P., Jr. [1941]. A general moment problem. *Am. J. Math.* **63**, 361–370.

Boas, R. P., Jr. [1952]. Integrability along a line for a class of entire functions. *Trans. Am. Math. Soc.* **73**, 191–197.

Boas, R. P., Jr. [1954]. "Entire Functions." Academic Press, New York.

Boas, R. P., Jr. [1955]. Isomorphism between H^p and L^p. *Amer. J. Math.* **77**, 655–656.

Boas, R. P., Jr. [1972]. Summation formulas and band-limited signals. *Tohoku Math. J.* **24**, 121–125.

Boas, R. P., Jr., and Pollard, H. [1948]. The multiplicative completion of sets of functions. *Bull. Am. Math. Soc.* **54**, 518–522.

Bochner, S. [1964]. Interpolation of general bounded and of almost periodic sequences by functions of stratified exponential type. *Proc. Natl. Acad. Sci. U.S.A.* **51**, 164–168.

Bočkarev, S. V. [1974]. A basis in the space of functions that are continuous in the closed disk and analytic in its interior. *Sov. Math. Dokl.* **15**, 1195–1198.

Bohr, H. [1925a]. Zur Theorie der fastperiodischen Funktionen. *Acta Math.* **45**, 29–127.

Bohr, H. [1925b]. Zur Theorie der fastperiodischen Funktionen. *Acta Math.* **46**, 101–214.

Bohr, H. [1926]. Zur Theorie der fastperiodischen Funktionen. *Acta Math.* **47**, 237–281.

Bohr, H. [1947]. "Almost Periodic Functions." Chelsea, Bronx, New York.

Borel, E. [1897]. Sur les zéros des fonctions entières. *Acta Math.* **20**, 357–396.

Borel, E. [1921]. "Leçons sur les fonctions entières," Deuxième édition revue et augmentée d'une note de M. G. Valiron. Gauthier-Villars, Paris.

Bourgain, J. [1988]. A remark on the uncertainty principle for Hilbertian bases. *J. Funct. Anal.* **79**, 136–143.

Bourgin, D. G. [1946]. A class of sequences of functions. *Trans. Am. Math. Soc.* **60**, 478–518.

Bratiščev, A. V., and Korobeĭnik, Ju. F. [1976]. The multiple interpolation problem in the space of entire functions of given proximate order. *Math. USSR-Izv.* **10**, 1049–1074.

Brauer, F. [1964]. The completeness of biorthogonal systems. *Mich. Math. J.* **11**, 379–383.

Byrnes, J. S. [1972]. Functions which multiply bases. *Bull. London Math. Soc.* **4**, 330–332.

Carleman, T. [1922]. Über die Approximation analytischer Funktionen durch lineare Aggregate von vorgegebene Potenzen. *Ark. Mat. Astron. Fys.* **17**(9), 1–30.

Carleson, L. [1958]. An interpolation problem for bounded analytic functions. *Am. J. Math.* **80**, 921–930.

Carleson, L. [1966]. On convergence and growth of partial sums of Fourier series. *Acta Math.* **116**, 135–157.

Carleson, L. [1980]. An explicit unconditional basis in H^1. *Bull. des Sci. Math.* **104**, 405–416.

Cartwright, M. L. [1935]. On certain integral functions of order 1 and mean type. *Proc. Cambridge Philos. Soc.* **31**, 347–350.

Cartwright, M. L. [1956]. "Integral Functions." Cambridge Univ. Press, London.

Casazza, P. G. [1997]. Characterizing Hilbert space frames with the subframe property. *Illinois J. Math.* **41**, 648–666.

Casazza, P. G. [1998]. Every frame is a sum of three (but not two) orthonormal bases–and other frame representations. *J. Fourier Anal. Appl.* **4**, 727–732.

Casazza, P. G. [1999]. Local theory of frames and Schauder bases for Hilbert space. *Illinois J. Math.* **43**, 291–306.

Casazza, P. G. [2000]. The art of frame theory. *Taiwanese J. Math.* **4**, 129–201.

Casazza, P. G., and Christensen, O. [1996]. Hilbert space frames containing a Riesz basis and Banach spaces which have no subspace isomorphic to c_0. *J. Math. Anal. Appl.* **202**, 940–950.

Casazza, P. G., and Christensen, O. [1998a]. Frames containing a Riesz basis and preservation of this property under perturbations. *Siam J. Math. Anal.* **29**, 266–278.

Casazza, P. G., and Christensen, O. [1998b]. Frames and Schauder bases. *Approximation Theory*, 133–139, *Monogr. Textbooks Pure Appl. Math.* **212**, Dekker, New York.

Casazza, P. G., and Christensen, O. [2000]. Approximation of the inverse frame operator and applications to Gabor frames. *J. Approx. Theory* **103**, 338–356.

Casazza, P. G., and Kalton, N. [1999]. Generalizing the Paley–Wiener perturbation theory for Banch spaces, *Proc. Amer. Math. Soc.* **127**, 519–527.

Casazza, P. G., Christensen, O., and Janssen, A. J. E. M. [1999]. Classifying tight Weyl-Heisenberg frames. *The functional and harmonic analysis of wavelets and frames (San Antonio, TX, 1999)*, 131–148, *Contemp. Math.* **247**, Amer. Math. Soc., Providence.

Casazza, P. G., Han, D., and Larson, D. R. [1999]. Frames for Banach spaces. *The functional and harmonic analysis of wavelets and frames (San Antonio, TX, 1999)*, 149–182, *Contemp. Math.* **247**, Amer. Math. Soc., Providence.

Chistyakov, G., and Lyubarskiĭ, Y. [1997]. Random perturbations of exponential Riesz bases in $L^2(-\pi, \pi)$. *Ann. Inst. Fourier (Grenoble)* **47**, 201–255.

Christensen, O. [1993]. Frames and the projection method. *Appl. Comput. Harm. Anal.* **1**, 50–53.

Christensen, O. [1995a]. A Paley–Wiener theorem for frames. *Proc. Amer. Math. Soc.* **123**, 2199–2201.

Christensen, O. [1995b]. Frame perturbations. *Proc. Amer. Math. Soc.* **123**, 1217–1220.

Christensen, O. [1996a]. Frames containing a Riesz basis and approximation of the frame coefficients using finite-dimensional methods. *J. Math. Anal. and Appl.* **199**, 256–270.

Christensen, O. [1996b]. Moment problems and stability results for frames with applications to irregular sampling and Gabor frames. *Appl. Comput. Harm. Anal.* **3**, 82–86.

Christensen, O. [1997]. Perturbations of frames and applications to Gabor frames. *In* "Gabor Analysis and Applications: Theory and Applications," pp. 193–209, H. G. Feichtinger and T. Strohmer, editors, Birkhäuser, Boston.

Christensen, O. [2000]. Finite-dimensional approximation of the inverse frame operator. *J. Fourier Anal. Appl.* **6**, 79–91.

Christensen, O., and Heil, C. [1997]. Perturbation of Banach frames and atomic decompositions. *Math. Nachr.* **185**, 33–47.

Christensen, O., Deng, B., and Heil, C. [1999]. Density of Gabor frames. *Appl. Comput. Harm. Anal.* **7**, 292–304.

Chui, C. K. [1992a]. "Wavelets: A Tutorial in Theory and Applications." Academic Press, New York.

Chui, C. K. [1992b]. "An Introduction to Wavelets." Academic Press, New York.

Chui, C. K., and He, W. [2000]. Compactly supported tight frames associated with refinable functions. *Appl. Comput. Harm. Anal.* **8**, 293–319.

Chui, C. K., and Shi, X. [1993a]. Bessel sequences and affine frames. *Appl. Comput. Harm. Anal.* **1**, 29–49.

Chui, C. K., and Shi, X. [1993b]. Inequalities of Littlewood-Paley type for frames and wavelets. *SIAM J. Math. Anal.* **24**, 263–277.

Chui, C. K., and Shi, X. [1996]. On stability bounds of perturbed multivariate trigonometric systems. *Appl. Comput. Harm. Anal.* **3**, 283–287.

Ciesielski, Z. [1963]. Properties of the orthonormal Franklin system. *Stud. Math.* **23**, 141–157.

Ciesielski, Z. [1966]. Properties of the orthonormal Franklin system. II. *Stud. Math.* **27**, 289–323.

Ciesielski, Z. [1986]. Bases in function spaces. *Approximation theory, V* (College Station, Texas, 1986), 31–54, Academic Press, Boston.

Ciesielski, Z., and Domsta, J. [1972]. Construction of an orthonormal basis in $C^m(I^d)$ and $W_p^m(I^d)$. *Stud. Math.* **41**, 211–224.

Ciesielski, Z., Simon, P., and Sjölin, P. [1977]. Equivalence of Haar and Franklin bases in L_p spaces. *Stud. Math.* **60**, 195–210.

Clarkson, J. A., and Erdös, P. [1943]. Approximation by polynomials. *Duke Math. J.* **10**, 5–11.

Cook, T. A. [1970]. On normalized Schauder bases. *Ann. Math. Mon.* **77**, 167.

Cossar, J. [1975/76]. Mean-square convergence of non-harmonic trigonometrical series. *Proc. R. Soc. Edinburgh*, 75A, **24**, 297–323.

Dalzell, D. P. [1945]. On the completeness of a series of normal orthogonal functions. *J. London Math. Soc.* **20**, 87–93.

Daubechies, I. [1992]. "Ten Lectures on Wavelets." SIAM, Philadelphia.

Daubechies, I., and Grossmann, A. [1988]. Frames in the Bargmann space of entire functions. *Comm. Pure Appl. Math.* **41**, 151–164.

Daubechies, I., and Janssen, A. [1993]. Two theorems on lattice expansions. *IEEE Trans. Inform. Theory* **39**, 3–6.

Daubechies, I., Grossmann, A., and Meyer, Y. [1986]. Painless nonorthogonal expansions. *J. Math. Phys.* **27**, 1271–1283.

Davie, A. M. [1973]. The approximation problem for Banach spaces. *Bull. London Math. Soc.* **5**, 261–266.

Davis, P. J. [1963]. "Interpolation and Approximation." Ginn (Blaisdell), New York.

Davis, W. J. [1969]. Basis preserving maps. *Proc. Am. Math. Soc.* **22**, 34–36.

de Branges, L. [1968]. "Hilbert Spaces of Entire Functions." Prentice-Hall, Englewood Cliffs.

Deutsch, F., and Morris, P. [1969]. On simultaneous approximation and interpolation which preserves the norm. *J. Approx. Theory* **2**, 355–373.

Diestel, J. [1975]. "Geometry of Banach Spaces — Selected Topics," Lecture Notes in Mathematics, No. 485. Springer-Verlag, Berlin and New York.

Dragilev, M. M., Zaharjuta, V. P., and Korobeĭnik, J. F. [1974]. A dual connection between certain questions of basis theory and interpolation theory. *Sov. Math. Dokl.* **15**, 533–537.

Duffin, R. J., and Eachus, J. J. [1942]. Some notes on an expansion theorem of Paley and Wiener. *Bull. Am. Math. Soc.* **48**, 850–855.

Duffin, R. J., and Schaeffer, A. C. [1938]. Some properties of functions of exponential type. *Bull. Am. Math. Soc.* **44**, 236–240.

Duffin, R. J., and Schaeffer, A. C. [1945]. Power series with bounded coefficients. *Am. J. Math.* **67**, 141–154.

Duffin, R. J., and Schaeffer, A. C. [1952]. A class of nonharmonic Fourier series. *Trans. Am. Math. Soc.* **72**, 341–366.

Dunford, N., and Schwartz, J. T. [1958]. "Linear Operators. Part I: General Theory." Wiley (Interscience), New York.

Dunford, N., and Schwartz, J. T. [1963]. "Linear Operators. Part II: Spectral Theory." Wiley (Interscience), New York.

Duren, P. L. [1970]. "Theory of H^p Spaces." Academic Press, New York.

Duren, P. L., and Williams, D. L. [1972]. Interpolation problems in function spaces. *J. Funct. Anal.* **9**, 75–86.
Dvoretzky, A., and Rogers, C. A. [1950]. Absolute and unconditional convergence in normed linear spaces. *Proc. Natl. Acad. Sci. U.S.A.* **36**, 192–197.
Earl, J. P. [1970]. On the interpolation of bounded sequences by bounded functions. *J. London Math. Soc.* **2**, 544–548.
Elsner, J. [1971]. Zulässige Abänderungen von Exponentialsystemen im $L^p(-A, A)$. *Math. Z.* **120**, 211–220.
Enflo, P. [1973]. A counterexample to the approximation property in Banach spaces. *Acta Math.* **130**, 309–317.
Eoff, C. [1995]. The discrete nature of the Paley-Wiener spaces. *Proc. Amer. Math. Soc.* **123**, 505–512.
Fattorini, H. O. [1977]. Estimates for sequences biorthogonal to certain complex exponentials and boundary control of the wave equation. *In* "Lecture Notes in Control and Information Sciences," Vol. 2, pp. 111–124.
Fattorini, H. O., and Russell, D. L. [1974]. Uniform bounds on biorthogonal functions for real exponentials and applications to the control theory of parabolic equations. *Quart. Appl. Math.* **32**, 45–69.
Favier, S. J., and Zalik, R. A. [1995]. On the stability of frames and Riesz bases. *Appl. Comput. Harm. Anal.* **2**, 160–173.
Favier, S. J., and Zalik, R. A. [1997]. Frames and Riesz bases: a short survey. *Wavelet theory and harmonic analysis in applied sciences (Buenos Aires, 1995)*, 93–117, *Appl. Numer. Harmon. Anal.*, Birkhäuser, Boston.
Feichtinger, H. [1990]. Coherent frames and irregular sampling. *Recent advances in Fourier analysis and its applications (Il Ciocco, 1989)*, 427–440, *NATO Adv. Sci. Inst. Ser. C Math. Phys. Sci.* **315**, Kluwer Acad. Publ., Dordrecht.
Feichtinger, H., and Gröchenig, K. [1992]. Irregular sampling theorems and series expansions of band-limited functions. *J. Math. Anal. Appl.* **167**, 530–556.
Feichtinger, H., and Gröchenig, K. [1994]. Theory and practice of irregular sampling. In "Wavelets: Mathematics and Applications," J. J. Benedetto and M. W. Frazier, editors, pp. 305–363.
Fejér, L., and Riesz, F. [1921]. Über einige functionentheoretische Ungleichungen. *Math. Z.* **11**, 305–314.
Feller, W. [1966]. "An Introduction to Probability Theory and its Applications," Vol. II. Wiley, New York.
Feller, W. [1968]. On Müntz' theorem and completely monotone functions. *Am. Math. Mon.* **75**, 342–350.
Flornes, K., Lyubarskiĭ, Y., and Seip, K. [1999]. A direct interpolation method for irregular sampling. *Appl. Comput. Harm. Anal.* **7**, 305–314.
Freund, E., and Petzold, J. [1984]. Nonharmonic Fourier series: a formalism for analyzing signals. *Elektron. Inf. Kybernet* **20**, 575–592.

Fujii, N., Nakamura, A., and Redheffer, R. [1999]. On the excess of sets of complex exponentials. *Proc. Amer. Math. Soc.* **127**, 1815–1818.

Gabardo, J-P. [1993]. Weighted tight frames of exponentials on a finite interval. *Monatsh. Math.* **116**, 197–229.

Gabardo, J-P. [1995]. Tight frames of polynomials and the truncated trigonometric moment problem. *J. Fourier Anal. Appl.* **1**, 249–279.

Gabor, D. [1946]. Theory of communication. *J. Inst. Electr. Eng., London* **93** (III), 429–457.

Gelbaum, B. R. [1950]. Expansions in Banach spaces. *Duke Math. J.* **17**, 187–196.

Gelbaum, B. R. [1951]. A nonabsolute basis for Hilbert space. *Proc. Am. Math. Soc.* **2**, 720–721.

Gelfand, I. M. [1951]. Remark on the work of N. K. Bari "Biorthogonal systems and bases in Hilbert space." *Učen. Zap. Mosk. Gos. Univ. 148, Mat.* **4**, 224–225.

Giroux, A., Rahman, Q. I., and Schmeisser, G. [1979]. On Bernstein's inequality. *Can. J. Math.* **31**, 347–353.

Goffman, C. [1964]. Remark on a problem of Lusin. *Acta Math.* **111**, 63–72.

Gohberg, I. C., and Krein, M. G. [1969]. "Introduction to the Theory of Linear Nonselfadjoint Operators," Translations of Mathematical Monographs, Vol. 18., Am. Math. Soc., Providence.

Golovin, V. D. [1964]. Biorthogonal expansions in linear combinations of exponential functions in L^2. *Zap. Har'kov. Gos. Univ. i Har'kov. Mat. Obšč (4)* **30**, 18–29.

Gosselin, R. P., and Neuwirth, J. H. [1968/69]. On Paley–Wiener bases. *J. Math. Mech.* **18**, 871–879.

Goupillaud, P., Grossmann, A., and Morlet, J. [1984/85]. Cycle-octave and related transforms in seismic signal analysis, *Geoexploration* **23**, 85–102.

Graves, L. M. [1941]. Some general approximation theorems. *Ann. Math.* **42**, 281–292.

Grinblium, M. M. [1948]. On a property of a basis. *Dokl. Akad. Nauk SSSR* **59**, 9–11.

Gripenberg, G. [1993a]. Wavelet bases in $L^p(R)$. *Studia Math.* **106**, 175–187.

Gripenberg, G. [1993b]. Unconditional bases of wavelets for Sobolev spaces. *SIAM J. Math. Anal.* **24**, 1030–1042.

Gröchenig, K. [1991]. Describing functions: atomic decompositions versus frames. *Monatsh. Math.* **112**, 1–42.

Gröchenig, K., and Razafinjatova, H. [1996]. On Landau's necessary density conditions for sampling and interpolation of band-limited functions. *J. London Math. Soc.* **54**, 557–565.

Gröchenig, K., and Ron, A. [1998]. Tight compactly supported wavelet frames of arbitrarily high smoothness. *Proc. Amer. Math. Soc.* **126**, 1101–1107.

Gröchenig, K., Heil, C., and Walnut, D. [2000]. Nonperiodic sampling and the local three squares theorem. *Arkiv. Mat.* **38**, 77–92.

Grossmann, A., and Morlet, J. [1984]. Decomposition of Hardy functions into square integrable wavelets of constant shape. *SIAM J. Math. Anal.* **15**, 723–736.

Grothendieck, A. [1955]. Produits tensoriels topologiques et espaces nucléaires. *Mem. Am. Math. Soc.* **16**, 1–191.

Gurariĭ, V. I., and Gurariĭ, N. I. [1971]. Bases in uniformly convex and uniformly flattened Banach spaces. *Math. USSR-Izv.* **5**, 220–225.

Gurariĭ, V. I., and Meletidi, M. A. [1970]. Stability of completeness of sequences in Banach spaces. *Bull. Acad. Pol. Sci., Ser. Sci. Math. Astron. Phys.* **18**, 533–536.

Gurevich, L. A. [1941]. Sur une propriété de la base dans l'espace de Hilbert. *Dokl. Akad. Nauk SSSR* **30**, 289–291.

Hadamard, J. [1893]. Etude sur les propriétés des fonctions entières et en particulier d'une fonction considerée par Riemann. *J. Math. pures et appliquées* **9**, 171–215.

Hahn, H. [1927]. Über lineare Gleichungssysteme in linearen Räumen. *J. Reine Angew. Math.* **157**, 214–229.

Halmos, P. R. [1951]. "Introduction to Hilbert Space and the Theory of Spectral Multiplicity." Chelsea, Bronx, New York.

Halmos, P. R. [1967]. "A Hilbert Space Problem Book." Van Nostrand, New York.

Hamburger, H. [1920]. Über eine Erweiterung des Stieltjesschen Momentenproblems. *Math. Ann.* **81**, 235–319.

Hammersley, J. M. [1953]. A nonharmonic Fourier series. *Acta Math.* **89**, 243–260.

Hansen, S. W. [1991]. Bounds on functions biorthogonal to sets of complex exponentials; control of damped elastic systems. *J. Math. Anal. Appl.* **158**, 487–508.

Hardy, G. H. [1949]. "Divergent Series." Oxford Univ. Press (Clarendon), London.

Hardy, G. H., Littlewood, J. E., and Pólya, G. [1952]. "Inequalities," 2nd ed. Cambridge Univ. Press, London.

Hausdorff, F. [1923]. Momentprobleme für ein endliches Intervall. *Math. Z.* **16**, 220–248.

Hedenmalm, H., Richter, S., and Seip, K. [1996]. Interpolating sequences and invariant subspaces of given index in the Bergman spaces. *J. Reine Angew. Math.* **477**, 13–30.

Heil, C. E. [1990]. Wavelets and frames. *Signal processing, Part I*, 147–160, *IMA Vol. Math. Appl.* **22**, Springer, New York.

Heil, C. E., and Walnut, D. F. [1989]. Continuous and discrete wavelet transforms. *SIAM Review* **31**, 628–666.

Heller, W. [1992]. Complex analysis and frames in sampling theory. *Probabilistic and stochastic methods in analysis with applications (Il Ciocco,*

1991), 101–116, *NATO Adv. Sci. Inst. Ser. C Math. Phys. Sci.* **372**, Kluwer Acad. Publ., Dordrecht.

Hernández, E., and Weiss, G. [1996]. "A First Course on Wavelets." CRC Press, Boca Raton.

Higgins, J. R. [1977]. "Completeness and Basis Properties of Sets of Special Functions." Cambridge University Press, Cambridge.

Higgins, J. R. [1985]. Five short stories about the cardinal series. *Bull. Am. Math. Soc.* **12**, 45–89.

Higgins, J. R. [1994]. Sampling theory for Paley–Wiener spaces in the Riesz basis setting. *Proc. R. Irish Acad.* **94A**, 219–236.

Hilding, S. [1948]. Note on completeness theorems of Paley–Wiener type. *Annals of Math.* **49**, 953–955.

Hille, E. [1972]. Introduction to general theory of reproducing kernels. *Rocky Mount. J. Math.* **2**, 321–368.

Hoffman, K. [1962]. "Banach Spaces of Analytic Functions." Prentice-Hall, Englewood Cliffs.

Holland, A. S. B. [1973]. "Introduction to the Theory of Entire Functions." Academic Press, New York.

Holub, J. R. [1987]. On perturbations of bases and basic sequences. *Rev. Roumaine Math. Pures Appl.* **32**, 611–616.

Holub, J. R. [1994]. Pre-frame operators, Besselian frames, and near-Riesz bases in Hilbert spaces. *Proc. Amer. Math. Soc.* **122**, 779–785.

Hruščev, S. V. [1979]. Perturbation theorems for bases of exponentials and Muckenhoupt's condition. *Sov. Math. Dokl.* **20**, 665–669.

Hruščev, S. V. [1987]. Unconditional bases in $L^2(0, a)$. *Proc. Amer. Math. Soc.* **99**, 651–656.

Hruščev, S. V., Nikolskiĭ, N. K., and Pavlov, B. S. [1981]. Unconditional bases of reproducing kernels. *Lecture Notes in Mathematics*, volume 864, pp. 214–235. Springer, Berlin.

Hunt, R. A., Muckenhoupt, B., and Wheeden, R. L. [1973]. Weighted norm inequalities for the conjugate function and Hilbert transform. *Trans. Amer. Math. Soc.* **176**, 227–251.

Ingham, A. E. [1936]. Some trigonometrical inequalities with applications to the theory of series. *Math. Z.* **41**, 367–379.

Ivanov, S. A. [1996]. Nonharmonic Fourier series in the Sobolev spaces of positive fractional orders. *New Zealand J. Math.* **25**, 39–46.

Jaffard, S. [1991]. A density criterion for frames of complex exponentials. *Mich. Math. J.* **38**, 339–348.

Jaffard, S., and Young, R. M. [1998]. A representation theorem for Schauder bases in Hilbert space. *Proc. Amer. Math. Soc.* **126**, 553–560.

Jaffard, S., Tucsnak, M., and Zuazua, E. [1997]. On a theorem of Ingham. *J. Fourier Anal. Appl.* **3**, 577–582.

James, R. C. [1945]. Orthogonality in normed linear spaces. *Duke Math. J.* **12**, 291–302.

James, R. C. [1950]. Bases and reflexivity of Banach spaces. *Ann. Math.* **52**, 518–527.

Janssen, A. J. E. M. [1995]. Duality and biorthogonality for Weyl-Heisenberg frames. *J. Fourier Anal. Appl.* **1**, 403–436.

Janssen, A. J. E. M. [1996]. Some counterexamples in the theory of Weyl-Heisenberg frames. *IEEE Trans. Inform. Theory* **42**, 621–623.

Jensen, J. L. W. V. [1899]. Sur un nouvel et important théorème de la théorie des fonctions. *Acta Math.* **22**, 359–364.

Jing, Z. [1999]. On the stability of wavelet and Gabor frames (Riesz bases). *J. Fourier Anal. Appl.* **5**, 105–125.

Johnson, W. B., Rosenthal, H. P., and Zippin, M. [1971]. On bases, finite-dimensional decompositions and weaker structures in Banach spaces. *Isr. J. Math.* **9**, 488–506.

Kac, M. [1941]. Convergence and divergence of non-harmonic gap series. *Duke Math. J.* **8**, 541–545.

Kadec, M. I. [1964]. The exact value of the Paley–Wiener constant. *Sov. Math. Dokl.* **5**, 559–561.

Kahane, J. P. [1958]. Sur la totalité des suites d'exponentielles imaginaires. *Ann. Inst. Fourier (Grenoble)* **8**, 273–275.

Kahane, J. P. [1966]. "Travaux de Beurling et Malliavin." Benjamin, New York.

Kaiser, G. [1994]. Deformations of Gabor frames. *J. Math. Physics* **35**, 1372–1376.

Karlin, S. [1948]. Bases in Banach spaces. *Duke Math. J.* **15**, 971–985.

Kato, T. [1976]. "Perturbation Theory for Linear Operators." Springer–Verlag, Berlin and New York.

Katsnelson, V. É. [1971]. Exponential bases in L^2. *Funct. Anal. Appl.* **5**, 31–38.

Kim, H. O., and Lim, J. K. [1997]. New characterizations of Riesz bases. *Appl. Comput. Harm. Anal.* **4**, 222–229.

Knopp, K. [1951]. "Theory and Application of Infinite Series," 2nd ed. Blackie, Glasgow and London.

Kober, H. [1943]. On the approximation to integrable functions by integral functions. *Trans. Amer. Math. Soc.* **54**, 70–82.

Koosis, P. [1960]. Sur la totalité des systèmes d'exponentielles imaginaires. *C. R. Acad. Sci.* **250**, 2102–2103.

Koosis, P. [1979]. Proof of the Beurling–Malliavin theorem by duality and harmonic estimation. *Harmonic Analysis in Euclidean Spaces, Proc. Symp. Pure Math.* **XXXV** (Part 1). Amer. Math. Soc., Providence.

Korevaar, J. [1947]. A characterization of the submanifold of $C[a, b]$ spanned by the sequence $\{x^{n_k}\}$. *Indag. Math.* **9**, 360–368.

Korobeĭnik, Ju. F. [1976]. Unconditional bases in a Hilbert space. *Math. Notes* **19**, 153–157.

Korobeĭnik, Ju. F. [1979]. The moment problem, interpolation, and basicity. *Math. USSR-Izv.* **13**, 277–306.

Köthe, G. [1936]. Das Trägheitsgesetz der quadratischen Formen im Hilbertschen Raum. *Math. Z.* **41**, 137–152.

Krein, M. G., and Liusternik, L. A. [1948]. Functional analysis. *In* "Thirty Years of Mathematics in the USSR (1917–1947)," pp. 608–672. Gos. Izdat. Tekhn.-Teor. Lit., Moscow-Leningrad.

Krein, M. G., and Nudelman, A. A. [1977]. "The Markov Moment Problem and Extremal Problems," Translations of Mathematical Monographs, Vol. 50. Am. Math. Soc., Providence.

Krein, M. G., Milman, D. P., and Rutman, M. A. [1940]. On a property of the basis in Banach space. *Zap. Mat. T-va (Kharkov)* **16**, 106–108.

Krotov, V. G. [1974]. Representation of measurable functions by series in the Faber-Schauder system, and universal series. *Sov. Math. Dokl.* **15**, 351–355.

Kwapien, S., and Pelczynski, A. [1970]. The main triangle projection in matrix spaces and its applications. *Studia Math.* **34**, 43–68.

Landau, H. J. [1964]. A sparse regular sequence of exponentials closed on large sets. *Bull. Am. Math. Soc.* **70**, 566–569.

Landau, H. J. [1967]. Necessary density conditions for sampling and interpolation of certain entire functions. *Acta Math.* **117**, 37–52.

Lawton, W. [1990]. Tight frames of compactly supported wavelets. *J. Math. Phys.* **31**, 1898–1901.

Lawton, W. [1991]. Necessary and sufficient conditions for constructing orthonormal wavelet bases, *J. Math. Phys.* **32**, 57–61.

Lerer, L. E. [1969]. The stability of bases of locally convex spaces. *Sov. Math. Dokl.* **10**, 24–28.

Levin, B. Ja. [1961]. On bases of exponential functions in L^2. *Zap. Har'kov. Gos. Univ. i Har'kov. Mat. Obšč. (4)* **27**, 39–48.

Levin, B. Ja. [1964]. "Distribution of Zeros of Entire Functions," Translations of Mathematical Monographs, Vol. 5. Am. Math. Soc., Providence.

Levin, B. Ja. [1969]. Interpolation by entire functions of exponential type. *Proc. Phys.-Technol. Inst. Low Temp., Acad. Sci. Ukr. SSR, Math. Phys. Funct. Anal.* No. 1, pp. 136–146.

Levin, B. Ja. [1996]. "Lectures on Entire Functions." Am. Math. Soc., Providence.

Levin, B. Ja., and Lyubarskiĭ, J. I. [1974]. Interpolation by entire functions of exponential type and its application to expansions in series of exponentials. *Funct. Anal. Appl.* **8**, 172–174.

Levin, B. Ja., and Lyubarskiĭ, J. I. [1975]. Interpolation by means of special classes of entire functions and related expansions in series of exponentials. *Math. USSR-Izv.* **9**, 621–662.

Levin, B. Ja., and Ostrovskiĭ, I. V. [1980]. On small perturbations of the set of zeros of functions of sine type. *Math. USSR Izvestija* **14**, 79–101.

Levinson, N. [1935]. On the closure of $\{e^{i\lambda_n x}\}$ and integral functions. *Proc. Cambridge Philos. Soc.* **31**, 335–346.

Levinson, N. [1936]. On the closure of $\{e^{i\lambda_n x}\}$. *Duke Math. J.* **2**, 511–516.

Levinson, N. [1940]. "Gap and Density Theorems," Am. Math. Soc. Colloq. Publ., Vol. 26. Am. Math. Soc., New York.

Li, S. [1995]. On general frame decompositions. *Numer. Funct. Anal. Optim.* **16**, 1181–1191.

Lin, B., and Singer, I. [1971]. On conditional bases of l^2. *Commentat. Math. Pr. Mat.* **15**, 135–139.

Lindenstrauss, J., and Pelczynski, A. [1968]. Absolutely summing operators in L_p-spaces and their applications. *Stud. Math.* **29**, 275–326.

Lindenstrauss, J., and Pelczynski, A. [1971]. Contributions to the theory of the classical Banach spaces. *J. Funct. Anal.* **8**, 225–249.

Lindenstrauss, J., and Tzafriri, L. [1977]. "Classical Banach Spaces I." Springer-Verlag, Berlin and New York.

Lindenstrauss, J., and Zippin, M. [1969]. Banach spaces with a unique unconditional basis. *J. Funct. Anal.* **3**, 115–125.

Lindner, A. M. [1999]. On lower bounds of exponential frames. *J. Fourier Anal. Appl.* **5**, 185–192.

Lorch, E. R. [1939]. Bicontinuous linear transformations in certain vector spaces. *Bull. Am. Math. Soc.* **45**, 564–569.

Lorentz, G. G. [1953]. "Bernstein Polynomials." Univ. of Toronto Press, Toronto.

Low, F. [1985]. *Complete sets of wave packets.* In "A Passion for Physics–Essays in Honor of Geoffrey Chew," pp. 17–22, World Scientific, Singapore.

Luxemburg, W. A. J. [1976]. Müntz-Szász type approximation results and the Paley–Wiener theorem. *In* "Approximation Theory, II" (G. G. Lorentz, C. K. Chui, and L. L. Schumaker, eds.), pp. 437–448. Academic Press, New York.

Luxemburg, W. A. J., and Korevaar, J. [1971]. Entire functions and Müntz-Szász type approximation. *Trans. Am. Math. Soc.* **157**, 23–37.

Lyubarskiĭ, Y. I. [1992]. Frames in the Bargmann space of entire functions. Entire and subharmonic functions, pp. 167–180, Adv. Soviet Math. **11**, Amer. Math. Soc., Providence.

Lyubarskiĭ, Y. I., and Seip, K. [1997]. Complete interpolating sequences for Paley–Wiener spaces and Muckenhoupt's (A_p) condition. *Revista Mat. Iber.* **13**, 361–376.

Mallat, S. G. [1989]. Multiresolution approximations and wavelet orthonormal bases for $L^2(R)$. *Trans. Amer. Math. Soc.* **315**, 69–87.

Mallat, S. G. [1999]. "A Wavelet Tour of Signal Processing," 2nd ed. Academic Press, New York.

Marcinkiewicz, J. [1937]. Quelques théorèmes sur les séries orthogonales. *Ann. Soc. Pol. Math.* **16**, 84–96.

Markushevich, A. I. [1965]. "Theory of Functions of a Complex Variable," Vol. 2. Prentice-Hall, Englewood Cliffs.

Markushevich, A. I. [1966]. "Entire Functions." Elsevier, Amsterdam.

Marti, J. T. [1969]. "Introduction to the Theory of Bases." Springer-Verlag, Berlin and New York.

Martirosjan, V. M. [1979]. Closure and basicity of certain biorthogonal systems and the solution of the multiple interpolation problem in $H_p[\alpha; \omega]$. *Sov. Math. Dokl.* **20**, 260–263.

Maurey, B. [1980]. Isomorphismes entre espaces H^1. *Acta Math.* **145**, 79–120.

Meletidi, M. A. [1972]. Complete systems and bases in the spaces C and L_p. *Sov. Math. Dokl.* **13**, 1372–1376.

Meyer, Y. [1986]. "Principe d'incertitude, bases Hilbertiennes et algèbres d'opérateurs." Séminaires Bourbaki **662**, Paris.

Meyer, Y. [1992]. "Wavelets and Operators." Cambridge University Press, Cambridge.

Meyer, Y. [1993]. "Wavelets: Algorithms and Applications." SIAM, Philadelphia.

Minkin, A. M. [1992]. Reflections of exponents and unconditional bases of exponentials. *St. Petersberg Math. J.* **3**, 1043–1068.

Morlet, J. [1983]. Sampling theory and wave propogation. *In* "NATO ASI Series," volume 1, pp. 233–261. Issues in acoustic signal/image processing and recognition. C. H. Chen, ed. Springer-Verlag, Berlin.

Morlet, J., Arens, G., Fourgeau, I., and Giard, D. [1982]. Wave propogation and sampling theory. *Geophysics* **47**, 203–236.

Müntz, C. H. [1914]. "Über den Approximationssatz von Weierstrass," pp. 303–312. H. A. Schwarz Festschr., Berlin.

Narukawa, K., and Suzuki, T. [1986]. Nonharmonic Fourier series and its applications. *Appl. Math. Optim.* **14**, 249–264.

Natanson, I. P. [1965]. "Constructive Function Theory," Vol. 2. Ungar, New York.

Nehari, Z. [1952]. "Conformal Mapping." McGraw-Hill, New York.

Neville, C. W. [1977]. A short proof of an inequality of Carleson's. *Proc. Am. Math. Soc.* **65**, 131–132.

Nikolskiĭ, N. K. [1980]. Bases of exponentials and the values of reproducing kernels. *Sov. Math. Dokl.* **21**, 937–941.

O'Hara, P. J. [1973]. Another proof of Bernstein's theorem. *Am. Math. Mon.* **80**, 673–674.

Orlicz, W. [1933]. Über unbedingte Konvergenz in Funktionenräumen I. *Stud. Math.* **4**, 33–37.

Ortega-Cerdà, J., and Seip, K. [1999]. Multipliers for entire functions and an interpolation problem of Beurling. *J. Functional Anal.* **162**, 400–415.

Oskolkov, V. [1978]. Some bases in spaces of regular functions and their application to interpolation. *Math. USSR-Sb.* **34**, 215–234.

Paley, R. E. A. C., and Wiener, N. [1934]. "Fourier Transforms in the Complex Domain," *Am. Math. Soc. Colloq. Publ.*, Vol. 19. *Am. Math. Soc.*, New York.

Pavlov, B. S. [1979]. Basicity of an exponential system and Muckenhoupt's condition. *Sov. Math. Dokl.* **20**, 655–659.

Pelczynski, A. [1961]. On the impossibility of embedding of the space L in certain Banach spaces. *Colloq. Math.* **8**, 199–203.

Pelczynski, A. [1977]. "Banach Spaces of Analytic Functions and Absolutely Summing Operators," Regional Conference Series in Mathematics, No. 30. Am. Math. Soc., Providence.

Pelczynski, A., and Singer, I. [1964]. On non-equivalent bases and conditional bases in Banach spaces. *Stud. Math.* **25**, 5–25.

Peterson, D. R. [1974]. The excess of sets of complex exponentials. *Proc. Am. Math. Soc.* **44**, 321–325.

Phragmén, E., and Lindelöf, E. [1908]. Sur une extension d'un principe classique de l'analyse et sur quelques propriétés des fonctions monogènes dans le voisinage d'un point singulier. *Acta Math.* **31**, 381–406.

Plancherel, M., and Pólya, G. [1938]. Fonctions entières et intégrales de Fourier multiples (Seconde partie). *Commentat. Math. Helv.* **10**, 110–163.

Pollard, H. [1944]. Completeness theorems of Paley–Wiener type. *Annals of Math.* **4**, 738–739.

Pollard, H., and Shisha, O. [1972]. Variations on the binomial series. *Am. Math. Mon.* **79**, 495–499.

Pólya, G. [1929]. Untersuchungen über Lücken und Singularitäten von Potenzreihen. *Math. Z.* **29**, 549–640.

Pólya, G. [1931]. Problem 108. *Jahresber. Dtsch. Math. Ver.* **40**, Part 2, 81.

Pólya, G., and Szegö, G. [1925]. "Aufgaben und Lehrsätze aus der Analysis," Vols. I and II. Springer-Verlag, Berlin and New York.

Povzner, A. [1949]. On the completeness of the sequence of functions $\{e^{i\lambda_n t}\}$ in $L^2(-\pi, \pi)$. *Dokl. Akad. Nauk SSSR* **64**, 163–166.

Redheffer, R. M. [1953]. On a theorem of Plancherel and Pólya. *Pac. J. Math.* **3**, 823–835.

Redheffer, R. M. [1954]. On even entire functions with zeros having a density. *Trans. Am. Math. Soc.* **77**, 32–61.

Redheffer, R. M. [1957]. Ganze Funktionen und Vollständigkeit. *Oesterr. Akad. Wiss.* **6**, 96–99.

Redheffer, R. M. [1968]. Elementary remarks on completeness. *Duke Math. J.* **35**, 103–116.

Redheffer, R. M. [1977]. Completeness of sets of complex exponentials. *Adv. Math.* **24**, 1–62.

Redheffer, R. M., and Young, R. M. [1983]. Completeness and basis properties of complex exponentials. *Trans. Am. Math. Soc.* **277**, 93–111.

Retherford, J. R., and Holub, J. R. [1971]. The stability of bases in Banach and Hilbert spaces. *J. Reine Angew. Math.* **246**, 136–146.

Riesz, F., and Sz.-Nagy, B. [1955]. "Functional Analysis." Ungar, New York.

Ron, A., and Shen, Z. [1995]. Frames and stable bases for shift-invariant subspaces of $L_2(R^d)$. *Can. Math. J.* **47**, 1051–1094.

Rosenbaum, J. T. [1967]. Simultaneous interpolation in H^2. *Mich. Math. J.* **14**, 65–70.

Rosenbaum, J. T. [1968]. Simultaneous interpolation in H_2, II. *Pac. J. Math.* **27**, 607–610.

Royden, H. L. [1968]. "Real Analysis," 2nd ed. Macmillan, New York.

Ruckle, W. H. [1972]. The extent of the sequence space associated with a basis. *Can. Math. J.* **24**, 636–641.

Rudin, W. [1966]. "Real and Complex Analysis." McGraw-Hill, New York.

Russell, D. L. [1967]. Nonharmonic Fourier series in the control theory of distributed parameter systems. *J. Math. Anal. Appl.* **18**, 542–560.

Russell, D. L. [1982]. On exponential bases for the Sobolev spaces over an interval. *J. Math. Anal. Appl.* **87**, 528–550.

Schäfke, F. W. [1949]. Über einige unendliche lineare Gleichungssysteme. *Math. Nachr.* **3**, 40–58.

Schauder, J. [1927]. Zur theorie stetiger Abbildungen in Funkionalräumen. *Math. Z.* **26**, 47–65, 417–431.

Schonefeld, S. [1972]. Schauder bases in the Banach spaces $C^k(T^q)$. *Trans. Am. Math. Soc.* **165**, 309–318.

Schuster, A., and Seip, K. [1998]. A Carleson-type condition for interpolation in Bergman spaces. *J. Reine Angew. Math.* **497**, 223–233.

Schwartz, L. [1943]. Approximation d'une fonction quelconque par des sommes d'exponentielles imaginaires. *Ann. Fac. Sci. Toulouse* **6**, 111–176.

Schwartz, L. [1959]. "Etude des Sommes d'Exponentielles," Publications de l'Institut de Mathématique de l'Université de Strasbourg, 2nd ed., Vol. V. Hermann, Paris.

Sedleckiĭ, A. M. [1970]. On functions periodic in the mean. *Math. USSR-Izv.* **4**, 1406–1428.

Sedleckiĭ, A. M. [1971a]. Nonharmonic Fourier series. *Sib. Math. J.* **12**, 793–802.

Sedleckiĭ, A. M. [1971b]. Interpolation by entire functions of exponential type. *Tr. Mosk. Orden. Lenin. Energ. Inst.* No. 89, pp. 55–64.

Sedleckiĭ, A. M. [1972a]. On biorthogonal expansions in exponential functions. *Math. USSR-Izv.* **6**, 579–586.

Sedleckiĭ, A. M. [1972b]. Continuation periodic in the mean and bases of exponential functions in $L^p(-\pi, \pi)$. *Math. Notes* **12**, 455–458.

Sedleckiĭ, A. M. [1973]. Equivalent sequences in certain function spaces. *Izv. Vyssh. Uchebn. Zaved., Mat.* **134** (No. 7), 85–91.

Sedleckiĭ, A. M. [1974]. The stability of the completeness and minimality in L^2 of a system of exponential functions. *Math. Notes* **15**, 121–124.

Sedleckiĭ, A. M. [1975a]. Equiconvergence and equisummability of nonharmonic Fourier expansions with ordinary trigonometric series. *Math. Notes* **18**, 586–591.

Sedleckiĭ, A. M. [1975b]. Expansions in exponential functions. *Sib. Math. J.* **16**, 628–635.

Sedleckiĭ, A. M. [1977a]. On a class of biorthogonal expansions in exponential functions. *Math. USSR-Izv.* **11**, 375–395.

Sedleckiĭ, A. M. [1977b]. Excesses of systems of exponential functions. *Math. Notes* **22**, 941–947.

Sedleckiĭ, A. M. [1978]. On completeness of the systems $\{\exp(ix(n+ih_n))\}$. *Anal. Math.* **4**, 125–143.

Sedleckiĭ, A. M. [1982]. Biorthogonal expansions of functions in series of exponents on intervals of the real axis. *Russian Math. Surveys* **37**, 57–108.

Sedleckiĭ, A. M. [1985]. Purely imaginary perturbations of the exponents λ_n in the system $\{\exp(i\lambda_n t)\}$. *Siberian Math. J.* **26**, 597–603.

Seip, K. [1995]. On the connection between exponential bases and certain related sequences in $L^2(-\pi, \pi)$. *J. Functional Anal.* **130**, 131–160.

Seip, K. [1998]. Developments from nonharmonic Fourier series. *Documenta Mathematica*. Extra volume ICM, II, 713–722.

Seip, K. [2000]. On Gröchenig, Heil, and Walnut's proof of the local three squares theorem. *Ark. Mat.* **38**, 93–96.

Seip, K., and Ulanovskii, M. [1996]. Random exponential frames. *J. London Math. Soc.* **53**, 560–568.

Seip, K., and Ulanovskii, M. [1997]. The Beurling-Malliavin density of a random sequence. *Proc. Amer. Math. Soc.* **125**, 1745–1749.

Sengupta, A., and Karnick, H. [1978]. Analysis of the neutron slowing down equation. *J. Math. Phys.* **19**, 2563–2569.

Shah, S. M. [1941]. A note on the classification of integral functions. *Math. Student* **9**, 63–67.

Shannon, C. E. [1949]. Communication in the presence of noise. *Proc. I.R.E.* **37**, 10–21.

Shannon, C. E., and Weaver, W. [1964]. "The Mathematical Theory of Communication." Univ. of Illinois Press, Urbana.

Shapiro, H. S. [1971]. "Topics in Approximation Theory," Lecture Notes in Mathematics, Vol. 187. Springer-Verlag, Berlin and New York.

Shapiro, H. S., and Shields, A. L. [1961]. On some interpolation problems for analytic functions. *Am. J. Math.* **83**, 513–532.

Shohat, J. A., and Tamarkin, J. D. [1943]. "The Problem of Moments," Math. Surveys, Vol. 1. Am. Math. Soc., New York.

Singer, I. [1970]. "Bases in Banach Spaces I." Springer-Verlag, Berlin and New York.

Sjölin, P. [1977]. The Haar and Franklin systems are not equivalent bases in L^1. *Bull. Acad. Pol. Sci., Ser. Sci. Math. Astron. Phys.* **25**(11), 1099–1100.

Stepanoff, W. [1925]. Sur quelques généralisations des fonctions presque périodiques. *C. R. Acad. Sci.* **181**, 90–94.

Stieltjes, T. J. [1894]. Recherches sur les fractions continues. *Ann. Fac. Sci. Toulouse* **8**, J1–J122; **9**, A5–A47.

Stray, A. [1972]. Approximation and interpolation. *Pac. J. Math.* **40**, 463–475.

Su, W., and Zhou, X. [1999]. A sharper stability bound of Fourier frames. *J. Fourier Anal. Appl.* **5**, 67–71.

Sun, W., and Zhou, X. [1999]. On Kadec's $\frac{1}{4}$-theorem and the stability of Gabor frames. *Appl. Comput. Harm. Anal.* **7**, 239–242.

Švedenko, S. V. [1974]. Interpolation in certain Hilbert spaces of analytic functions. *Math. Notes* **15**, 56–61.

Švedenko, S. V., and Turku, H. [1975]. Interpolation in the class H^2 in a half-plane. *Sov. Math. Dokl.* **16**, 1547–1550.

Szász, O. [1916]. Über die Approximation stetiger Funktionen durch lineare Aggregate von Potenzen. *Math. Ann.* **77**, 482–496.

Szász, O. [1933]. Solution to problem 108. *Jahresber. Dtsch. Math. Ver.* **43**, Part 2, 20–23.

Szász, O. [1944]. "Introduction to the Theory of Divergent Series." University of Cincinnati, Cincinnati.

Szegö, G. [1939]. "Orthogonal Polynomials." *Am. Math. Soc.*, New York.

Sz.-Nagy, B. [1947]. Expansion theorems of Paley–Wiener type. *Duke Math. J.* **14**, 975–978.

Talaljan, F. A. [1972]. Rearrangements of series in a Hilbert space. *Math. Notes* **12**, 599–602.

Taylor, A. E. [1958]. "Introduction to Functional Analysis." Wiley, New York.

Taylor, B. A., and Williams, D. L. [1972]. Interpolation of l^q sequences by H^p functions. *Proc. Am. Math. Soc.* **34**, 181–186.

Titchmarsh, E. C. [1925]. A theorem on trigonometrical series. *Proc. London Math. Soc.* **XXIV**, xxii–xxiii.

Titchmarsh, E. C. [1926]. The zeros of certain integral functions. *Proc. London Math. Soc.* **25**, 283–302.

Titchmarsh, E. C. [1928]. A class of trigonometrical series. *J. London Math. Soc.* **3**, 300–304.

Titchmarsh, E. C. [1939]. "The Theory of Functions," 2nd ed. Oxford Univ. Press, London.

Titchmarsh, E. C. [1946]. "Eigenfunction Expansions Associated with Second-order Differential Equations." Clarendon Press, Oxford.

Titchmarsh, E. C. [1965]. On the mean convergence of trigonometrical series. *J. London Math. Soc.* **40**, 594–596 (addendum, **43**, 367).

Tolimiera, R., and Orr, R. [1995]. Poisson summation, the ambiguity function, and the theory of Weyl-Heisenberg frames. *J. Fourier Anal. Appl.* **1**, 233–247.

Ullrich, D. C. [1989]. Functions of exponential type and separated sequences. *Proc. Amer. Math. Soc.* **105**, 523–524.

Valiron, G. [1949]. "Lectures on the General Theory of Integral Functions." Chelsea, Bronx, New York.

Verblunsky, S. [1956]. On an expansion in exponential series. *Q. J. Math.* **7**, 231–240.

Vitali, G. [1921]. Sulla condizione di chiusura di un sistema di funzioni ortogonali. *Atti. R. Accad. Naz. Lincei, Rend. Cl. Sci. Fis., Mat. Nat.* **30**, 498–501.

Volkmer, H. [1995]. Frames of wavelets in Hardy spaces. *Analysis* **15**, 405–421.

Voss, J. J. [1999]. On discrete and continuous norms in Paley–Wiener spaces and consequences for exponential frames. *J. Fourier Anal. Appl.* **5**, 193–201.

Walker, W. [1987]. The separation of zeros for entire functions of exponential type. *J. Math. Anal. Appl.* **122**, 257–259.

Wallen, L. J. [1969]. A stability theorem for frames in Hilbert space. *Mich. Math. J.* **16**, 149–151.

Walsh, J. L. [1921]. A generalization of the Fourier cosine series. *Trans. Am. Math. Soc.* **22**, 230–239.

Walsh, J. L. [1935]. "Interpolation and Approximation by Rational Functions in the Complex Domain," *Am. Math. Soc. Colloq. Publ.*, Vol. 20. Am. Math. Soc., New York.

Weierstrass, K. [1876]. Zur Theorie der eindeutigen analytischen Funktionen. *Abh. K. Akad. Wiss.* 11–60 (see *Werke* **2**, 77–124 , 1895).

Weierstrass, K. [1885]. Über die analytische Darstellbarkeit sogennanter willkürlicher Funktionen einer reellen Veränderlichen. *Berl. Ber.* 633–640; 789–806 (see *Werke* **3**, 1–37, 1903).

Weierstrass, K. [1895]. "Mathematische Werke," 7 vols. [1894–1927]. Mayer & Müller, Berlin.

Whittaker, E. T. [1915]. On the functions which are represented by the expansions of the interpolation theory. *Proc. R. Soc. Edinburgh* **35**, 181–194.

Whittaker, J. M. [1935]. "Interpolatory Function Theory," Cambridge Tracts in Mathematics and Mathematical Physics, No. 33. Cambridge Univ. Press, London.

Wojtaszczyk, P. [1982]. The Franklin system is an unconditional basis in H^1. *Arkiv. für Mat.* **20**, 293–300.

Wojtaszczyk, P. [1991]. "Banach Spaces for Analysts." Cambridge University Press, Cambridge.

Wojtaszczyk, P. [1997]. "A Mathematical Introduction to Wavelets." Cambridge University Press, Cambridge.

Wolibner, W. [1951]. Sur un polynôme d'interpolation. *Colloq. Math.* **2**, 136–137.

Young, R. M. [1974a]. Interpolation in a classical Hilbert space of entire functions. *Trans. Am. Math. Soc.* **192**, 97–114.

Young, R. M. [1974b]. An extension of the Hausdoff–Young theorem. *Proc. Am. Math. Soc.* **45**, 235–236.

Young, R. M. [1974c]. Inequalities for a perturbation theorem of Paley and Wiener. *Proc. Amer. Math. Soc.* **43**, 320–322.

Young, R. M. [1975a]. A note on a trigonometric moment problem. *Proc. Am. Math. Soc.* **49**, 411–415.

Young, R. M. [1975b]. On perturbing bases of complex exponentials in $L^2(-\pi, \pi)$. *Proc. Am. Math. Soc.* **53**, 137–140.

Young, R. M. [1976a]. A perturbation theorem for complete sets of complex exponentials. *Proc. Am. Math. Soc.* **55**, 318–320.

Young, R. M. [1976b]. Interpolation for entire functions of exponential type and a related trigonometric moment problem. *Proc. Am. Math. Soc.* **56**, 239–242.

Young, R. M. [1976c]. Some stability theorems for nonharmonic Fourier series. *Proc. Am. Math. Soc.* **61**, 315–319.

Young, R. M. [1979]. On a completeness theorem of Paley and Wiener. *Proc. Am. Math. Soc.* **76**, 349–350.

Young, R. M. [1981]. On complete biorthogonal sequences. *Proc. Amer. Math. Soc.* **83**, 537–540.

Young, R. M. [1983]. On the pointwise convergence of a class of nonharmonic Fourier series. *Proc. Amer. Math. Soc.* **89**, 65–73.

Young, R. M. [1984]. On a theorem of Ingham on nonharmonic Fourier series. *Proc. Amer. Math. Soc.* **92**, 549–553.

Young, R. M. [1987]. On the stability of exponential bases in $L^2(-\pi, \pi)$. *Proc. Amer. Math. Soc.* **100**, 117–122.

Young, R. M. [1997]. Interpolation and frames in certain Banach spaces of entire functions. *J. Fourier Anal. Appl.* **3**, 639–645.

Young, R. M. [1998]. On a class of Riesz-Fischer sequences. *Proc. Amer. Math. Soc.* **126**, 1139–1142.

Zalik, R. A. [1999]. Riesz bases and multiresolution analyses. *Appl. Comput. Harm. Anal.* **7**, 315–331.

Zygmund, A. [1959]. "Trigonometric Series," 2nd ed. Cambridge Univ. Press, London.

List of Special Symbols

Author Index

Numbers in italics refer to pages on which the references are listed.

C

D

E

F

G

H

U

V

W

Y

Z

Subject Index

K

L

M

N

O

P

Q

R

S

T

U

V

W

Z